Climate Crisis Changed: The Intergovernmental Panel on Climate Change (IPCC) reports are deliberate science fiction

by

White, Wysmuller, Nelson, Beers, McMenemie

RoseDog Books
PITTSBURGH, PENNSYLVANIA 15238

The contents of this work, including, but not limited to, the accuracy of events, people, and places depicted; opinions expressed; permission to use previously published materials included; and any advice given or actions advocated are solely the responsibility of the author, who assumes all liability for said work and indemnifies the publisher against any claims stemming from publication of the work.

RoseDog Books
585 Alpha Drive, Suite 103
Pittsburgh, PA 15238
Visit our website at *www.rosedogbookstore.com*

ISBN: 979-8-89341-007-5
eISBN: 979-8-89341-506-3

Climate Crisis Changed

Cctruth.org

The Intergovernmental Panel On Climate Change

Reports are Deliberate Science fiction (IPCC).

High School Textbook

For General and Environmental Science

Dedication

This high school textbook is dedicated to the more than **thirty** doctoral level Scientists on our IPCC review committee who have fought so tirelessly against the political manipulation of science. At cctruth.org they are known affectionately as "the junk science slayers."

Dorrance Publishing Co., Lauriat Publishing.

Library of Congress ISBN

Cctruth.org

First Edition

White, Wysmuller, Beers, McMenemie, Nelson

TABLE OF CONTENTS

Forward

As far back as the 1990's, two of our Board members, Randy Beers and Dave White, took Dave's RV to Mount Hood Meadows Ski Resort every Friday evening in the Oregon winter. They'd open a single malt scotch bottle and sit around the fire debunking everything they'd heard about global warming and, later, climate change. Randy has a degree in physics and Dave is a chemical engineer with post-graduate study in advanced statistics. Randy would say things like the warming was due to sun spots and Dave would respond with something like atmospheric carbon dioxide doesn't freeze in the atmosphere.

As a result of all this collegial prognostication, in 2016 Dave launched Climate Change Truth, Inc. (cctruth.org), a 501(c)(3) nonprofit research corporation. When he realized that virtually nothing had been accomplished by controlling emissions of carbon dioxide to lower atmospheric carbon dioxide, he set about gathering all the data. His statistical analyses demonstrated conclusively that carbon dioxide emissions are not the cause of the rise in carbon dioxide levels. Then what is?

Atmospheric carbon dioxide is a binary system statistically. The two causal factors are carbon dioxide emissions and loss of photosynthesis. Since carbon dioxide emissions are clearly not the cause, then there must exist a loss of photosynthesis cause and solution.

This book explores the profound implications of these findings for climate change policy at every level of government. Moreover, it demonstrates the relative ease by which the climate crisis may be resolved and the important steps already underway to resolve it by collaboration with nations around the world. China, India, Pakistan, and Peru are leading the way. The solution is hiding in plain sight.

The book is designed as a high school textbook for sophomore students and above. Prerequisites are high school Statistics, Chemistry, Calculus, and General Science.

This textbook is appropriate for use in any

High School environmental or general science class.

Chapter 1

Statistical Analysis

Atmospheric carbon dioxide is a binary system statistically. A binary system consists of two possible causes for an effect. Binary statistical system has rules which guide the scientist toward the correct answer. One of the rules is that "the correlation coefficient must be greater than or equal to 0.9." This is in every statistics book. The correlation coefficient describes the closeness of two lines of data on a chart. The more parallel the lines, the higher the correlation coefficient. The number 1 represents perfectly parallel lines. When the correlation is less than 0.9, there is no cause and effect. So, scientists perform linear regression analysis by target to determine what the cause-and-effect variable correlates with.

Very large data sets are most easily analyzed as a binary system. Every large data collection must be viewed this way. Higher level college statistics courses teach that a very large data set must be viewed as a binary system whenever possible.

Scientists must first determine the two probable causes. Atmospheric carbon dioxide rise is a very large data set. It has been measured hourly since around 1960. Thus, we know that the binary system is emissions of carbon dioxide and loss of photosynthesis consumption of carbon dioxide.

Graph 1-1 is the well-known chart of atmospheric carbon dioxide rise vs. human emissions of carbon dioxide. Worldwide, this graph has not been updated since 2012 anywhere online except at cctruth.org. This graph misleads everyone to conclude that emissions are the cause of the rise. In spite of hundreds of personal contacts and protests since 2016 by cctruth.org, the scientists who publish this false picture refuse to change their stance. They continue to report this fallacy to world governments, who then proceed to make global climate change policy based on false data. These policies founded on false information affect people and businesses worldwide with economic hardship.

Graph 1-2 is the updated graph of human emissions vs. atmospheric carbon dioxide. Since 2014 the data shows a slope change from 0.6 gtur-1 to 0.3 gtyr-1. This is a 50% change in slope. Additionally, it shows the drop in worldwide emissions due to COVID and a rebound in 2022. However, this had zero effect on the rise of carbon dioxide.

Graph 1-3 is the 160 years of Troposphere Carbon Dioxide rise in the world. The current big change in slope started around 1950. We know the Emissions of carbon dioxide are not the whole cause of the rise. We need to find out how much of the rise is caused by Emissions of Carbon Dioxide. To do this we dray lines from the change in slope at 1950 to each say 10 ppm increase of the rise. Then calculate the correlation coefficient for each line. Then make a graph of correlation coefficient's and see where the peak is such as the graph 1-4. Graph 1-4 is Pearson's regression by target. Human emissions of carbon dioxide correlate to 363 ppm of the atmospheric carbon dioxide rise. Table 1-1 is the Pearson's regression by target data for graph 5-3.

The correlation coefficient is 0.63 and is thus not the cause and effect of carbon dioxide rise. We've known this since 2016. Why do you think this has never been shared with government policy makers?

After spending $2.8 trillion we have:

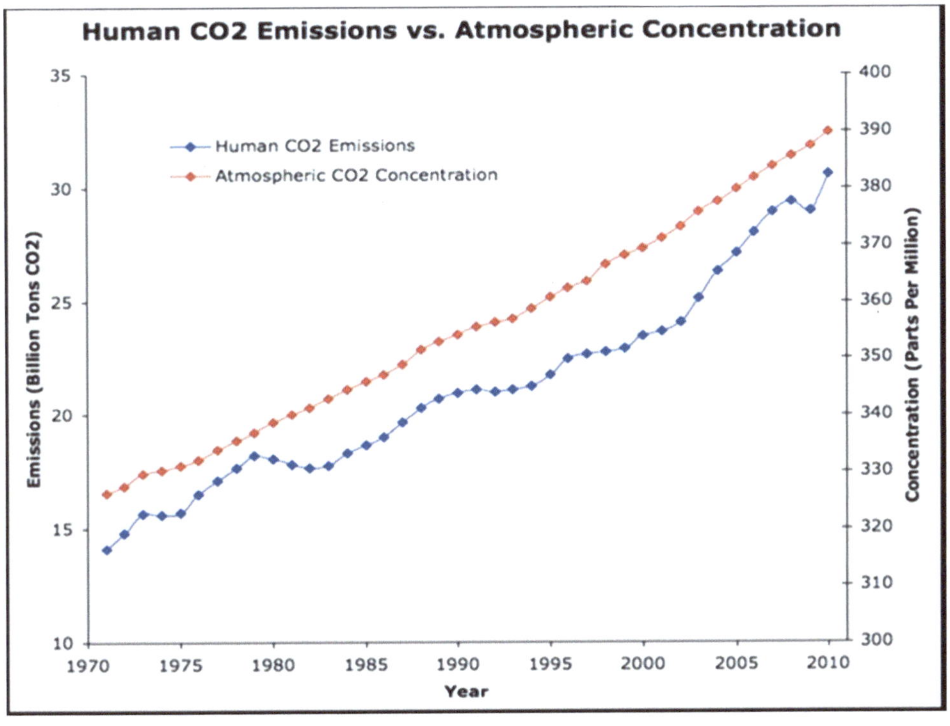

Graph 1-1 Mauna Loa Hawaii NOAA carbon dioxide rise.

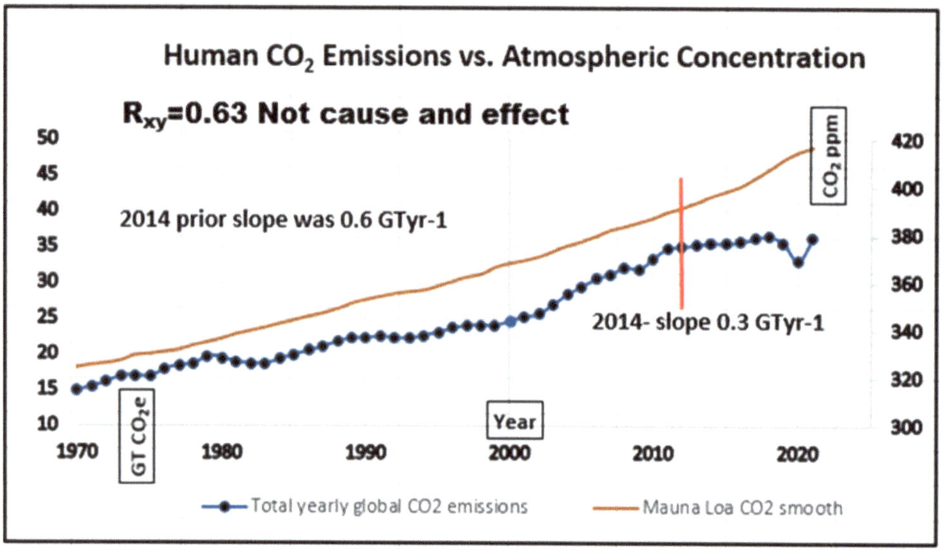

Graph 1-2. Updated every year graph of atmospheric CO_2 vs. emissions of CO_2.

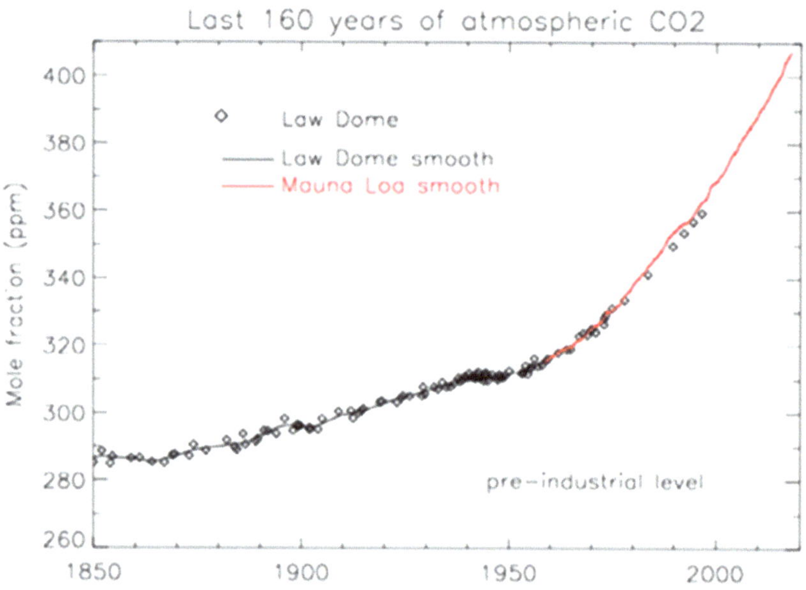

Graph 1-3 Last 160 years of Troposphere Carbon Dioxide

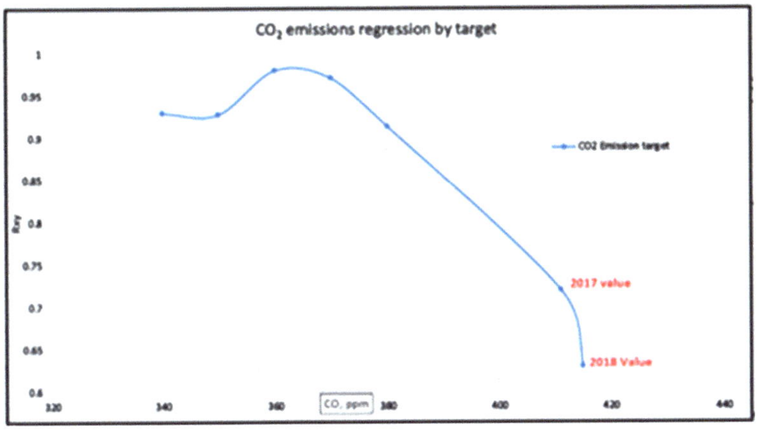

Graph 1.4 Pearson's regression by target

Pearson's Regression by Target for CO_2 emissions (all)		
Target (ppm)	Rxy	Comment
340	0.93	
350	0.928078	
360	0.98	
370	0.970839	
380	0.914433	
411	0.72	
415	0.64	2019 Value

$$r_{xy} = \frac{\sum_{i=1}^{n}(x_i - \bar{x})(y_i - \bar{y})}{\sqrt{\sum_{i=1}^{n}(x_i - \bar{x})^2 \sum_{i=1}^{n}(y_i - \bar{y})^2}}$$

http://cctruth.org/regression.xlxs

Table 1-1 Data from Pearson regression by target

Chapter 1 Questions:

1. What is a binary system?
2. Why should big data always use a binary system?
3. What is a multivariate system?

The Scientific Method

Science can prove or disprove virtually anything as long as the correct method is followed. Science can draw conclusions from data which are very reliable. Over time as more data accumulates, theories are proven with correct data analyses. The Scientific Method is the process by which unbiased scientific method is used to prove or disprove a theory. The scientific method starts with observation. We observe something we can't readily explain. The observation is based on data collection. From there we make a scientific guess about what is causing an effect based on observation. This is called the hypothesis. The hypothesis is a scientific inference based on observation as to the cause or answer to some scientific inquiry.

Data Collection

Correct experimental design is of extreme importance in constructing an experiment to collect data that proves or disproves a hypothesis. The experiment must be designed to collect the data needed to prove the hypothesis. See Chapter 11. The experimental design must also be designed to eliminate effects which are not part of the experiment.

Data collection must proceed in a manner so as to minimize miscellaneous effects. After enough data is collected, the hypothesis is proven or disproven. If proven, then additional data equal to or greater than the previous data set is needed to derive a scientific law. If the hypothesis is not proven, then the hypothesis must be changed to match the data. Then collect more data to prove the new hypothesis.

Scientific Law is a theory which has been proven by big data sets. Confirmation is needed. In addition, scientific law must be confirmed by other scientists performing the same experiments and achieving the same results. Scientific Law outside confirmation is still Scientific law. However, some confirmation may be false because the same experiment yielded a different conclusion. When this happens the scientific law is held back until more data is collected to prove or disprove it.

Chapter 2

Carbon Dioxide Equilibrium

White et al., J Earth Sci Clim Change 2022, 13:1

Journal of Earth Science & Climatic Change

Research Article Open Access

The Essential Role of Photosynthesis in Defining Net Zero Carbon Dioxide Emissions for Equilibrium Calculations

Dave White[1]*, Henry Ealy[2] and Kathrine Martin[3]

[1]*Chemical Engineer with Masters Level study in Statistics, Climate Change Truth Inc. USA*
[2]*Head of Energetic Health Institute and Operation Phoenix Rising, USA*
[3]*Research assistant in research department for Climate Change Truth, USA*

Abstract

In this research manuscript, the authors seek to answer four essential questions relative to the current climate change conversation now underway globally: (Q1) what is the numerically defined goal for annual Net Zero Carbon Dioxide Emissions in gigatonnes essential for global atmospheric homeostasis? (Q2) Why is atmospheric CO_2 rising even though recent data support that CO_2 emissions have the rate of rise lowered by 50% since 2014 globally? (Q3) Are CO_2 cap and trade policies the best immediate intervention, or does globally increasing photosynthesis offer a more rapid and better long-term solution to climate change? (Q4) What strategies can be employed to have the greatest positive impact over the upcoming crucial twelve-year period? Nothing absorbs carbon dioxide out of our atmosphere like photosynthesis, and therein lies the most under-discussed solution to the greatest problem of our time. A single hectare of healthy Amazon Rainforest can sequester up to 100 tons of CO_2yr-1 due to photosynthesis. And the fast-growing Empress Tree (Paulownia tomentosa) not only grows ten to twenty feet tall in its first year, but a single hectare of these trees can sequester up to 103 tons of CO_2yr-1 due to photosynthesis [1].

Prior to the Industrial Revolution and long before global deforestation devastated Earth's delicate atmospheric ecosystem, forests around the world are estimated to have consumed up to 400 billion tons of CO_2yr-1. As of 2019, that has been reduced dramatically as global forests consume less than 10 billion tons of CO_2yr-1 with photosynthesis [2].

Keywords: Photosynthesis; Carbon dioxide increase; Carbon dioxide scavenging; Climate change; Amazon rainforest

Significance Statement

The vast majority of climate experts agree that there has been a five- time increase in CO_2 emissions due to human related factors since 1870. While fossil fuel carbon have been confirmed to be approximately seven gigatons annually and does weigh in the climate change discussion, during this same period the photosynthetic consumption of carbon dioxide has been reduced by more than 97% due to incessant global deforestation. Historical forestry records indicate that prior to the 1900's; annual worldwide carbon dioxide consumption was estimated to have been around 400 gigatons due to photosynthesis. However, as of 2019, calculated estimates now have annual carbon dioxide consumption due to photosynthesis below ten gigatons.

(Q1) What Is the Numerical Goal for Annual Net Zero Carbon Emissions? (Net Zero CO_2e)

As a result of the global reduction in forest size, there has been a corresponding loss in photosynthesis and the natural carbon dioxide sequestration it creates. Estimates for global carbon dioxide sequestration have fallen from a minimum of 400 gigatons (healthy rain-forests and other forests) circa 1700 to as little as 10-12 gigatons annually, far below what is required to maintain global atmospheric homeostasis (Christine L. Goodale et al. 2002). The total photosynthesis in the Northern Hemisphere is less than 0.7 $gtCyr^{-1}$ (2.57 $gtCO_2yr^{-1}$) (Christine L. Goodale et al. 2002).

Currently, the burning of fossil fuels releases approximately (total worldwide CO_2 emissions) 36.8 gigatons of carbon dioxide each year. In order for humanity to restore the global atmospheric homeostasis essential for life, we must aggressively plant trees and regrow our beleaguered forests as the foundation of any comprehensive strategy for reversing climate change.

Deforestation has resulted in a massive global loss of carbon dioxide sequestration of more than 390 gigatons per year, that is simply no longer sustainable. Thus, we submit that the most pressing issue facing humanity in our climate change battle is not carbon dioxide emissions but, rather, the inability to absorb the carbon dioxide currently stranded in the atmosphere due to the significant reduction in photosynthesis. Our forests are the lungs of our planet. Without them, we simply cannot breathe.

The logic of the current climate science theory is based on the hypothesis that we can stop the dramatic shifts in weather patterns that threaten all life by achieving global atmospheric equilibrium as we remove excess carbon dioxide from the atmosphere. However, a major problem in this discussion of atmospheric equilibrium is that the phrase Net Zero Carbon Emissions (Net Zero CO_2e) has not been defined. What number, in gigatons, is the target goal that would give us atmospheric equilibrium? How much carbon dioxide needs to be consumed annually to get to the Net ZeroCO_2e target goal?

Productive scientific communication for the development of a unified global solution to the greatest challenge in human history depends on scientific data that have been reviewed and scrutinized for accuracy before agreement can be reached. In today's climate change conversations, some climate change scientists have suggested that the Net Zero CO_2e target goal is zero, meaning that every year we would need to absorb 36.8 gigatons of CO_2. Others simply use the term Net Zero CO_2e without having any numerical idea of what the target goal is and thus, how much carbon dioxide will need to be removed from the atmosphere each year in order to achieve equilibrium.

The definition of Net Zero CO_2e is, therefore, crucial to establishing a foundation for the entire climate change scientific community before it can develop practical strategies designed to stalemate the progression of the rise of atmospheric carbon dioxide and ultimately reverse its threatening course.

The author has calculated the Net Zero CO_2e to be 8.6 gigatons per year, and this research article will demonstrate in detail how that number was determined with data sets available in "References." As of 2019, the estimated global emission of carbon dioxide stands at 35.5 gigatons. Therefore, a Net Zero CO_2e of 8.6 gigatons means that in order for us to achieve atmospheric equilibrium and global environmental homeostasis, we must re-establish our ability to sequester 26.9 gigatons (26.9 billion tons) of CO_2 directly from the atmosphere each year.

Global reforestation is much more cost effective, time effective, and rapidly attainable than relying on industry and individual governments to curtail their pursuit of control and market share. After all, what is easier to do? (A) Plant trees, as Ethiopia has done with 350 million new saplings planted in twelve hours (NOAA weekly 2020), or (B) Reduce emissions of carbon dioxide, which has not worked to lower atmospheric carbon dioxide in the past. Additionally, reducing emissions won't work for the future because the average residence time for carbon dioxide in the future is 150 years (Schwartz, 2018).

Logic dictates that with a global catastrophe facing all humanity, now is the time for action with proven, natural, and trusted mechanisms for restoring global atmospheric equilibrium.

The Photosynthesis Equation

Photosynthesis is the biological process by which the plant cell organelle, known as a chloroplast, is able to "change" carbon dioxide into glucose to support its production of energy (ATP) for cellular function, including the replicated biotransformation of CO_2 into glucose. Carbon dioxide is literally the preferred fuel source for all plant life. With an abundance of it stranded in our atmosphere, there is more than enough fuel to support mass global reforestation. Here is the biochemical equation for photosynthetic sequestration of carbon dioxide from the atmosphere for your reference: $6\ CO_2 + 6\ H_2O\ \lambda\rightarrow C_6H_{12}O_6 + 6\ O_2$

How Is Annual CO_2 Consumption Lost Due to Lost Photosynthesis?

1. Globally, we lose at least 1 gigaton of CO_2 consumption annually due to city expansion into previously undeveloped lands. (Calculated by worldwide cities over 200k people Ha squared)

2. The IPPC Forestry Group estimates we lose 2-3 gigatons of CO_2 consumption annually from bio-mass burning due to forest fires, which excludes forest clearing for animal agriculture and wood commodity production.

3. Deforestation for animal agriculture and wood commodities in the Amazon Rainforest alone accounts for a minimal loss of 90 gt of CO_2 consumption annually (Brienen, R. et al. 2015).

According to the IPCC SR 1.5, the minimum carbon dioxide sequestration required is 45% of human carbon dioxide emissions by 2030 (Rogelj, J et al. 2018). This reduces our CO_2 emissions to 20 $GtCO_2\ yr^{-1}$ --in agreement with the statement of 25–30 $GtCO_2e\ yr^{-1}$ in chapter 2 of SR 1.5 However, there exists no reference to where the 25-30 GT $CO_2e\ yr^{-1}$ data came from. Planting two trees every year for every man, woman, and child on Earth each year for ten years will increase carbon sequestration from Earth's atmosphere by 1.4 gigatons each year and by 14 gigatons per year by the end of the first decade (see Northern Hemisphere Forest Photosynthesis below for explanation). We calculated the current worldwide photosynthesis from land and ocean surface, and it is 8.6 $GtCO_2e\ yr^{-1}$. Additionally, the total oxygen produced is 17.74 Gt yr^{-1} (E. Kintisch, 2017). Oxygen levels are declining. The "natural" carbon dioxide emissions are above the fossil fuel emissions. A 45% cut in human emissions will bring us to 20 Gt yr^{-1} and not to equilibrium. Following are the calculations for the Northern and Southern Hemispheres.

Forest Photosynthesis in the Northern hemisphere

Total forest hectares in the Northern Hemisphere are about 2050 million (CHRISTINE L. GOODALE et al. 2002). There are no rainforests in the Northern Hemisphere, equivalent to the Amazon Rainforest. The total photosynthesis in the Northern Hemisphere is less than 0.7 $GTCyr^{-1}$ (2.57 $GTCO_2yr^{-1}$) (CHRISTINE L. GOODALE et al. 2002). Increased biomass from the Earth warming and higher troposphere carbon dioxide concentrations do not produce enough photosynthesis to offset the massive deforestation around the globe. This is not what lowers the oscillation at Mauna Loa during the Northern Hemisphere summer.

Forest Photosynthesis in the Southern Hemisphere

The Southern Hemisphere (SH) has much less land than the Northern Hemisphere, but it does have rainforests. Unfortunately, the Amazon Rainforest has switched to become a massive CO_2 producer and oxygen sink (D. White, 2019), (Max Roser 2015), (Robert Scribbler 2020).

Because decay of plant material ingests oxygen and emits carbon dioxide, the decay contribution from the Amazon Rainforest is now ten to fifteen billion more tons of CO_2 annually with a corresponding loss of oxygen. This is derived from a conservative calculation of five tons of decay per acre per year.

Other research papers show that all tropical forests in the Southern Hemisphere (SH) have switched to become oxygen sinks and carbon dioxide producers. One such paper from A. Baccini et al. shows this clearly (A. Baccini Et al. 2017). In 2009, the tropical forests in the SH started to switch to oxygen sinks and carbon dioxide emitters due to organic decay. Anthropogenic forest degradation and biomass burning (forest fires and agricultural burning) also represent relevant contributions. The decay from the over two billion acres remaining of the Amazon Rainforest cancels out any other land- and ocean-based photosynthesis in the SH (see ocean photosynthesis below). This is because of its switching to an oxygen sink and Carbon Dioxide emitter (see figure 2 below and the massive carbon dioxide release from the Amazon--the white arrow points to this area) (Takahashi, Taro et al. 2008).

All tropical forests in the southern hemisphere have switched to be an oxygen sink and carbon dioxide producer due to organic decay (Takahashi, Taro et al. 2008). Certainly, if the Amazon had not switched, then the carbon dioxide around it would have a much lower concentration. Up to 90 gigatons (90 billion tons) CO_2 consumption is lost annually from Amazon Rainforest switching, 2.5 times our carbon dioxide emissions at 35.1 $gtCO_2yr^{-1}$. The worldwide massive deforestation explains why the average residence time of atmospheric carbon dioxide is increasing.

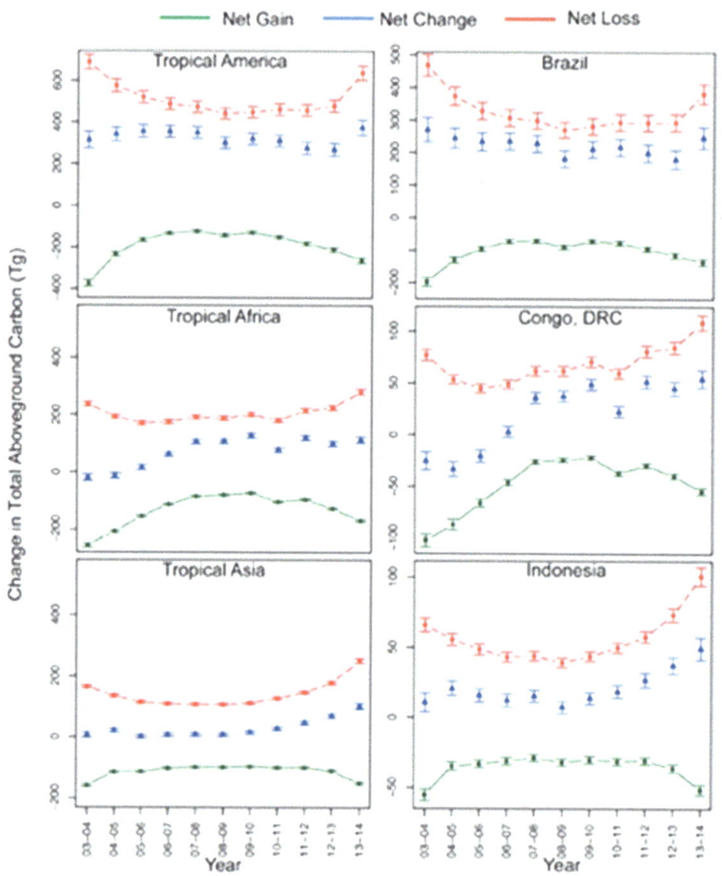

Figure 1

Figure 1. Most southern hemisphere tropical forests have switched to be an oxygen sink and carbon dioxide producer. Annual net change (95% confidence interval) in total carbon. Red lines indicate a loss in carbon; green lines indicate a gain. Blue lines reflect the difference between losses and gains. The vertical bars indicate the standard error of the change value. (A. Baccini Et al. 2017)

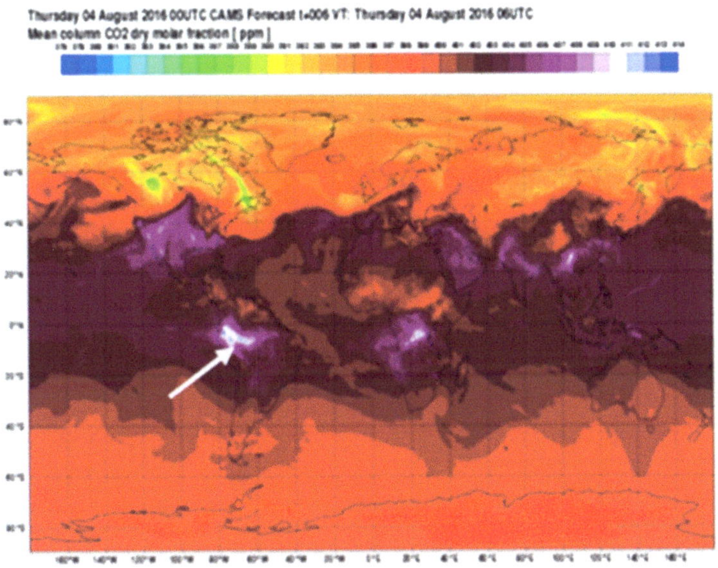

Figure 2. Amazon Rainforest Switched (Takahashi, Taro et al. 2008)

Oxygen Levels

Since the oxygen levels are decreasing, the photosynthesis must be less than that decrease (see Graph 1). The total worldwide oxygen needed is 39 Gt yr^{-1} (see Figure 3). The oxygen from declining photosynthesis and ocean outgassing produced is 17.74 GTO_2 yr^{-1}

Graph 1 Oxygen Levels Decreasing (Oxygen Levels 2019)

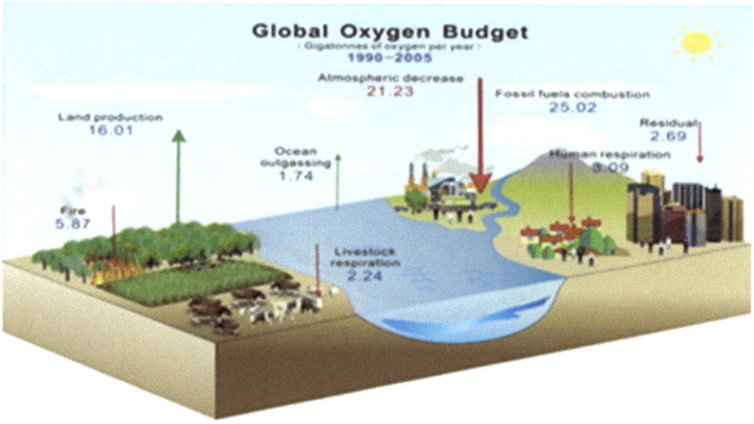

Figure 3. Worldwide Oxygen Budget (Jianping Huang et al 2018)

Regarding Q2, why is atmospheric CO_2 rising even though recent data support that CO_2 emissions show the rate of rise lowered by 50% since 2014 globally?

The logical reason that atmospheric CO_2 continues to rise in spite of CO_2 emissions rate of rise decreasing in half globally (see Figure 7) is that we do not have enough trees (primarily due to deforestation) to photo synthetically reabsorb carbon dioxide from the atmosphere. Additionally, the minimum residence time of atmospheric carbon dioxide is increasing (IPCC WG I 2003). As a result, the carbon that is still being emitted from all sources remains in the atmosphere and accumulates as a result of the 26.9 gigatons sequestering deficit gap.

As countries like China, India, Pakistan, and Ethiopia have shown, our future is dependent upon trees and our ability to redevelop nature's carbon dioxide sinks. Reforesting our planet can substantially increase global photosynthesis to first meet, and then exceed, the 28.2 gigatons that must be absorbed annually and buy us the time we need to address the myriad, additional companion issues like ocean clean-up, animal agriculture, and biodegradable plastic solutions.

Atmospheric carbon dioxide concentration is still increasing even though the emissions slope has decreased 60% (see Graph 2). Additionally, the rate of rise of atmospheric CO_2 is still increasing (see Graph 6). This confirms no effect on atmospheric CO_2 rise by slowing emissions.

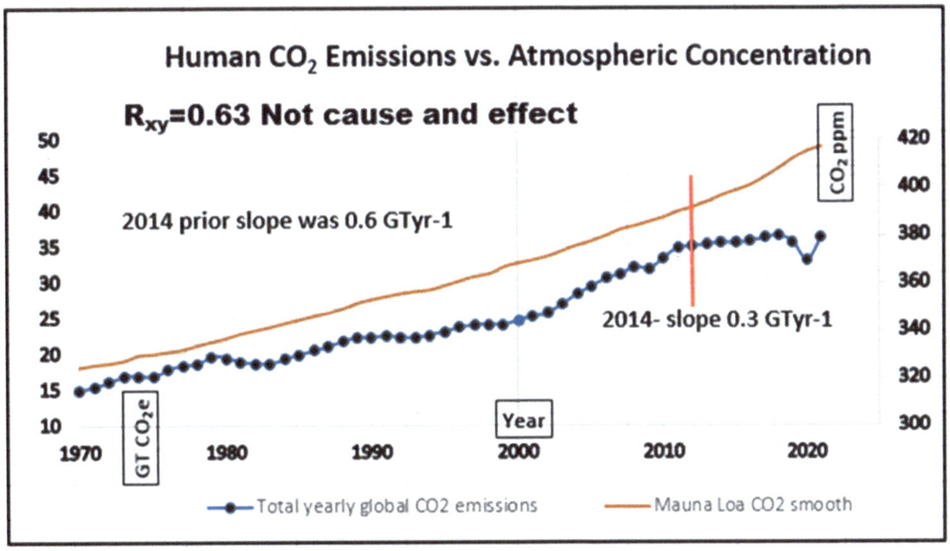

Graph 2. (D. White 2022)

Atmospheric CO_2 is a binary system statistically [6,18]. Two causes are CO_2 emissions and loss of photosynthesis. Each cause is multivariate. We have had mostly flat human emissions (0.3 GT/yr vs. 0.6 GT/yr) since 2014 (see Graph 2). However, atmospheric CO_2 is still going up and the rate of rise is increasing (Graph 8). In 2018, the Rxy correlation coefficient was 0.73 and not cause and effect. In 2021 it is now 0.63.

The lead author performed regression by target and CO_2 emissions correlate to 363 ppm of the rise in atmospheric carbon dioxide since 1950. (D. White, 2019).

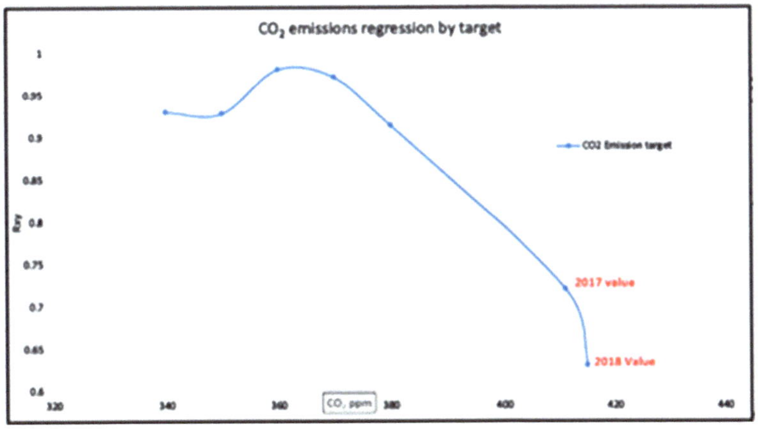

Graph 3. (D. White 2020)

Pearson's Regression by Target for CO_2 emissions (all)		
Target (ppm)	Rxy	Comment
340	0.93	
350	0.928078	
360	0.98	
370	0.970839	
380	0.914433	
411	0.72	
415	0.64	2019 Value

$$r_{xy} = \frac{\sum_{i=1}^{n}(x_i - \bar{x})(y_i - \bar{y})}{\sqrt{\sum_{i=1}^{n}(x_i - \bar{x})^2 \sum_{i=1}^{n}(y_i - \bar{y})^2}}$$

http://cctruth.org/regression.xlxs

Table 1. Pearson's Regression (D. White 2020)

Table 1 shows Pearson's regression by target starting at the inflection point (1950). Natural and human carbon dioxide emissions correlate to 363 ppm and are a contributor but not the cause of the rise of atmospheric carbon dioxide since 1957.

Oxygen Levels

Since the oxygen levels are decreasing, the photosynthesis must be less than that decrease (see Graph 1). The total worldwide oxygen needed is 39 Gt yr^{-1} (see Figure 3). The oxygen from declining photosynthesis and ocean outgassing produced is 17.74 GTO_2 yr^{-1}.

Effective CO_2 Emissions

Effective CO_2 emissions are emissions per year after subtracting the total photosynthesis consumption for the year. For example, the worldwide emissions for 2018 are 37.1 GT. The worldwide photosynthesis is 9.2 GT. Thus, the effective emissions would be 27.9 GT. Calculating it this way tells the actual effect. Instead of the emissions effect being at 135 ppm of atmospheric CO_2, it would be 81 ppm from Effective CO_2 Emissions (ECE). This correlates well with total emissions effect at 363 ppm.

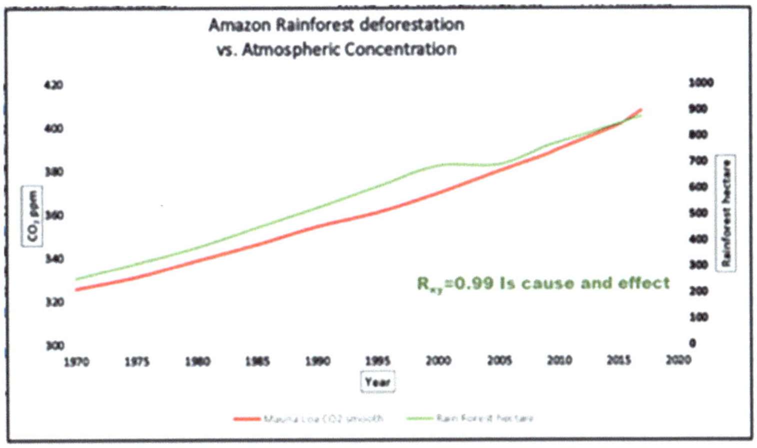

Graph 4. Rain forest deforestation vs. Atmospheric CO_2 rise.(D. White 2020)

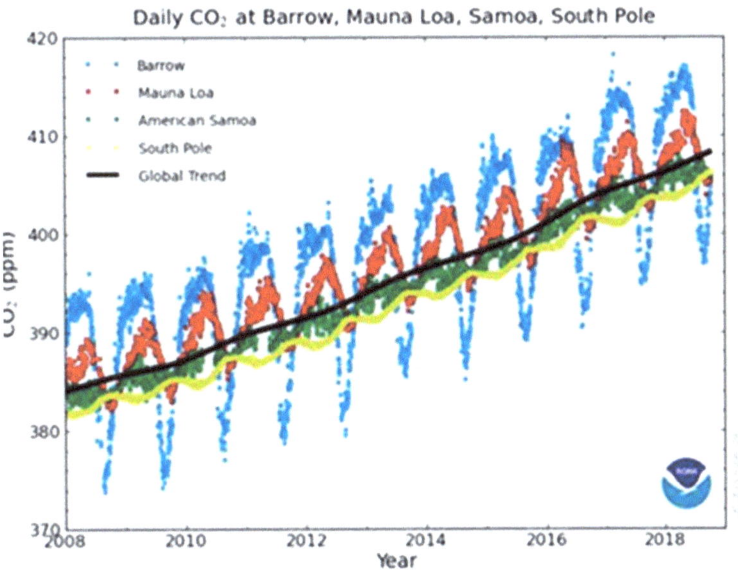

Graph 5. Mauna Loa cycles. (NOAA gr 2020)

Figure 4. Amazon Rainforest burning (Kenzie Mastroe 2019)

The oscillation at Mauna Loa starts as a very strong signal in South America and then fans out larger and larger until it reaches Barrow's Alaska. The countries in South America burn the Amazon Rainforest, the densest forest in the world, from October/November through May of the next year. Since 1950, an average of 30 million acres per year have been deforested and burned. So much CO_2 has been released that the trees and plants have grown too fast and died.

This massive decay is what caused the Amazon Rainforest to switch to an oxygen sink and carbon dioxide producer. Hundreds of manuscripts have been published on this. Currently, the Amazon output is 15 $GTyr^{-1}$ of CO_2. The switching of the Amazon Rainforest is a 0.99 cause and effect correlation to the rise in atmospheric carbon dioxide and loss of oxygen since 1957. The oscillation at Mauna Loa increases during the seven months of the deforestation and burning in South America's spring and summer (see figure 4). Then at the end of May, the deforestation stops, and the Mauna Loa carbon dioxide data recede.

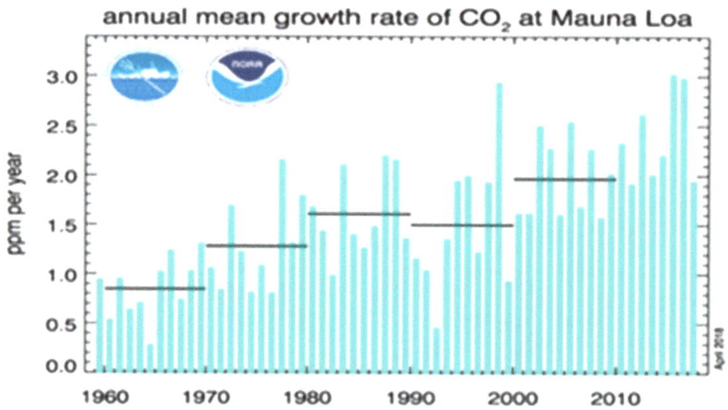

Graph 6. Mauna Loa CO_2 Rate of Rise (NOAA gr 2020)

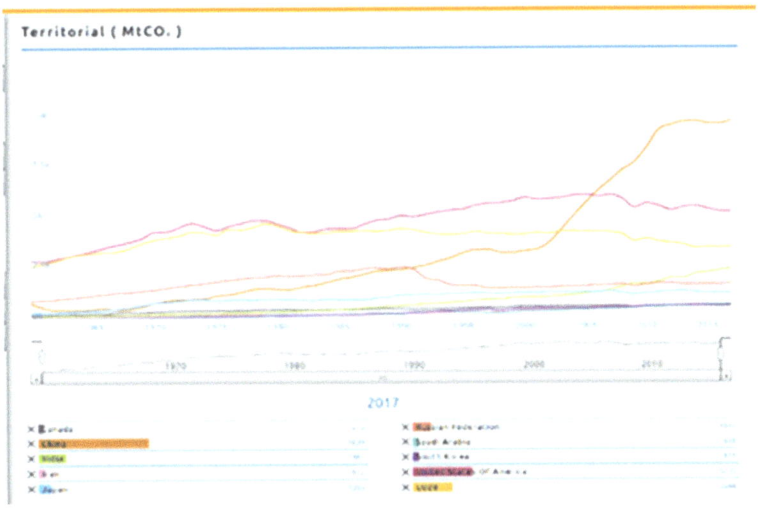

Figure 5. Global Carbon Emissions (Global Carbon Atlas 2020)

The rate of rise of atmospheric CO_2 is still increasing even though the global emissions have leveled off (see graph 6). All countries have lowered their carbon dioxide emissions except China and India (see figure 5). All pollutants are measured by mass and should be reported as such, not reported per capita. China is the worst polluter of carbon dioxide and has not reduced any emissions. The USA has been decreasing its CO_2 emissions since 2006, and Europe has been decreasing since 1990.

USA 2006: 6131 $MtCO_2$ and in 2018: 5270 $MtCO_2$ --a 15% decrease of carbon dioxide.
Europe 1990: 4479 $MtCO_2$ and in 2018: 3544 $MtCO_2$ --a 21% decrease of carbon dioxide.

Ocean Photosynthesis

The oceans in the Northern Hemisphere are currently producing six billion tons of photosynthesis consumption of carbon dioxide (Natalya Gallo 2014). This decline is for many reasons. However, if the oceans are a pump, the area without CO_2 is similar to the area with CO_2 (see figure 6). Therefore, the output and the input cancel each other. Ocean photosynthesis is declining because of a lowering of pH (Natalya Gallo 2014). Excess carbon dioxide in the oceans disassociate to carbonate ions. This causes less carbon dioxide available for photosynthesis. The oceans out gassing of oxygen is 1.74 gtyr-1 (see figure 3). Photosynthesis is 1:1 carbon to oxygen. Therefore, the total worldwide ocean consumption of carbon dioxide is 1.74 gtyr-1

Cruise Ship Effect

In 2017, 25 million passengers traveled on cruise ships. The ships with on-board sewage treatment plants are so few that their effect is negligible. Thus, with the average cruise lasting seven days, the total passenger days were 175 million (2017 data). Because people on cruise ships tend to eat more than usual, the wet weight of feces per person is maximum, at 1.5 kg (C. Rose, et al 2015). The excrement is 30% dry weight. Therefore, the mass is .3*1.5=0.45 kg.

Each passenger contributes a little over 0.45 kg of solid waste per day.

Therefore, there were 79,000 tons of solid waste in 2017. Most of this is long-chained hydrocarbons with 50 - 125 carbons each. These will be converted to CO_2 by the decay process. For CO_2 the molecular weight is 36 and carbon is 12. Thus, the weight increases by nine for each carbon converted. Consequently, 73 million tons of CO_2 were added from cruise ships in 2017. Since 1990, the decay of human waste dumped into the oceans from cruise ships, based on the reported number of passengers each year, has resulted in a total of 1.1 billion metric tons of carbon dioxide in the oceans. This decay over forty years has removed oxygen and produced carbon dioxide in the oceans.

If all the cruise ships were made to offload their passengers' waste products at ports and the rivers were cleaned of their pollution (e.g. the Petite Nèthe River near Antwerpen, polluted with horse manure), the ocean CO_2 would decrease. Only one port on the west coast of the USA has a facility to treat passengers' waste products, but no one is utilizing it. This port is in San Francisco. The person in charge reports that no vessel uses the facility to offload human waste. In addition, most ships dump as soon as they get out of the harbor.

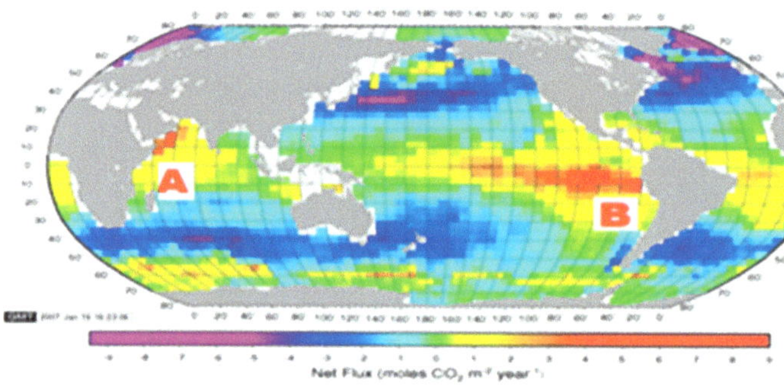

Figure 6. Takahashi 1997 Ocean Surface CO_2

Diatoms Effect

By assimilating the last fifteen years of satellite ocean chlorophyll in an established biogeochemical model, Cecile S. Rousseaux et al 2015 found that there are some significant changes in physical conditions, nutrients, and phytoplankton communities in the high latitudes. In the Northern Hemisphere, a shallowing of the mixed layer depth and a decline in nutrients affect differently the phytoplankton community, depending on the regions (Cecile S. Rousseaux et al 2015).

Diffusion of CO_2 in the Troposphere

Greenhouse gases, like all gases, diffuse until they are equidistant to each other at any given pressure and temperature combination. At STP (Standard Temperature and Pressure, 25C, 1 Atmosphere), CO_2 has the following diffusion coefficients:

In air: 16 mm^2/s
In water: 0.0016 mm^2/s (D. White 2020)

CO_2 is more likely to diffuse in the air than in the ocean (D. White 2020). The diffusion length in air (bulk troposphere) is 2 cm per month toward the exosphere (D. White 2020), (Ficks diffusion 2018). The ocean-air interface diffusion is 14.8 cm per day in the direction of the atmosphere. The driving force for diffusion is much greater in the direction of the exosphere (Machida et aL. 2002). Flux = 2 cm per month towards the exosphere (Ficks diffusion 2018). Eddy Diffusion and mixing of the atmosphere are stronger drivers for transport of CO_2 to higher altitudes (Hans R. Schneider et al. 1989).

The flux direction is correct. The flux number may or may not be correct. This is because of varying pressures and temperatures from the troposphere to the exosphere. In the future, I will perform a rigorous calculation to determine the accurate flux number. CO_2 that goes into the ocean is from any disturbance of ocean surface (e.g. hurricanes) that allows CO_2 to enter the ocean (Takahashi, Taro et al. 2008).

Most of the ocean's surface is at standard temperature and pressure at any time. Furthermore, the atmospheric winds distribute carbon dioxide evenly by latitude (see Graph 7 below). Except for temperature effect, every latitude has an equal chance to diffuse into the ocean. However, the concentrations in the flux graph are not close to what diffusion principles would indicate (see Figure 6 above). Takahashi et al. show southern hemisphere rates of change of ocean pCO_2 (see Table 3). Also see Figure 6. The most concentration of ocean pCO_2 is in the southern hemisphere. If this were from diffusion flux, one would naturally expect the flux rate to be larger toward the equator and smaller toward the south pole. Figure 8 shows no such dependency. The top row is closest to the south pole and it shows the highest rate! Furthermore, the PMEL website shows no difference in PCO2 concentration or ocean acidity for any link chosen (PMEL ocean acidity data). The only location which seems to show ph. decreasing is at Mauna Loa Hawaii.

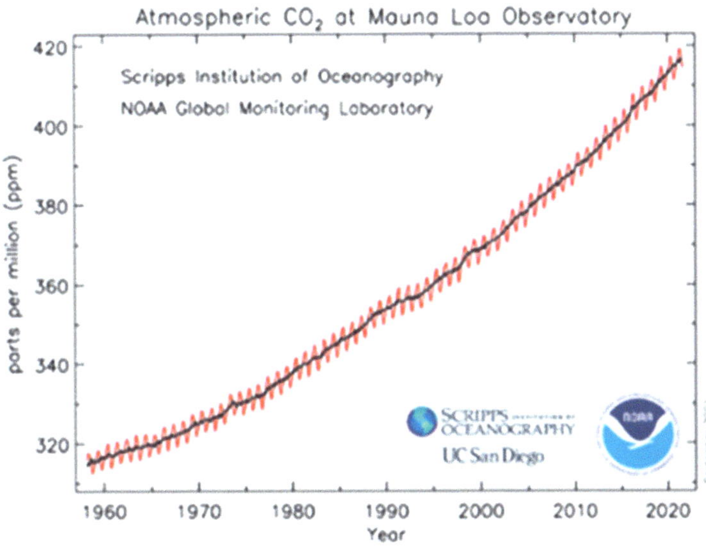

Figure 7. Mauna Loa CO2 rise since 1960. (9/24/2021)

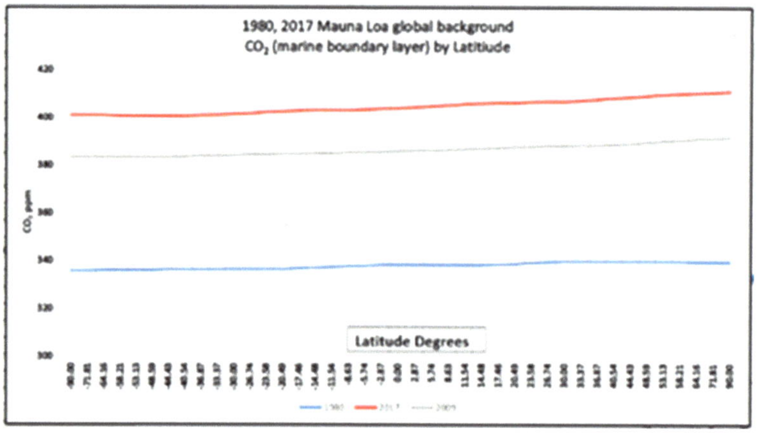

Graph 7. CO_2 mixed by atmospheric winds. (D. White, P. Tans (NOAA) 2017)

$$J = -D \cdot \frac{dc(x)}{dx} \qquad \text{(Unit: D: cm}^2\text{/sec;} \quad \text{J: number/cm}^2\text{/sec)}$$

Figure 8. Ficks First Law (Welty et al, 1984)

Mean rates of change in surface water pCO_2 and bulk water temperature of mixed layer (SST) estimated in six areas in the temperate South Pacific Ocean

Locations	Lat.	Long.	Data period	Rate of change		No. of [illegible]	Rate of change	
				pCO_2/SST (µatm y^{-1})	±σ (µatm y^{-1})		SST (°C y^{-1})	±σ (°C y^{-1})
Tahiti	15-20°S	135-145°W	1974-1997	2.00	0.39	16	0.033	0.034
Vanuatu	20-25°S	165-175°E	1984-2006	1.30	0.27	35	-0.048	0.018
New Caledonia	25-30°S	170-180°W	1976-2005	1.05	0.09	17	-0.013	0.030
Tasmania	43-48°S	140-150°E	1984-2004	1.83	0.56	18	-0.051	0.062
New Zealand	45-50°S	170-180°E	1976-2006	1.42	0.30	37	0.001	0.027
S. of Tasmania	50-55°S	140-150°E	1984-2002	1.61	0.30	12	-0.044	0.064
Mean				1.5±0.3	0.30	-	-0.02±0.05	0.04

The rates are computed using mean monthly values. The temperatures are measured concurrently with pCO_2.

Table 3. Diffusion by latitude in SH (Takahashi, Taro et al. 2008)

Worldwide

The total worldwide consumption of carbon dioxide is 2.6 (NH) +6(NH ocean) = 8.6 gigatonnes per year. Therefore, we cannot lower atmospheric carbon dioxide by working on emissions. It is a waste of resources because emissions are not the primary cause. Logically, 90% of human carbon dioxide emissions are from the NH (see Figure 9). The Mauna Loa carbon dioxide data peak in May each year, and then decline until rising again in November (see Figure 10).

The greater economic activity during the summertime in the NH produces greater amounts of carbon dioxide emissions. Then, in the fall each year, when there is less activity, and it is not cold yet, emissions decrease. In November, the temperature drops in the NH, and more fossil fuel for heating is consumed. This increases the carbon dioxide emissions. In April, it starts warming and carbon dioxide emissions decline (see graph 8). However, the Mauna Loa carbon dioxide data show that concentration decreases during the same time that human emissions increase in the summer. Furthermore, the total photosynthesis in the NH forests is 2.6 gigatonnes per year.

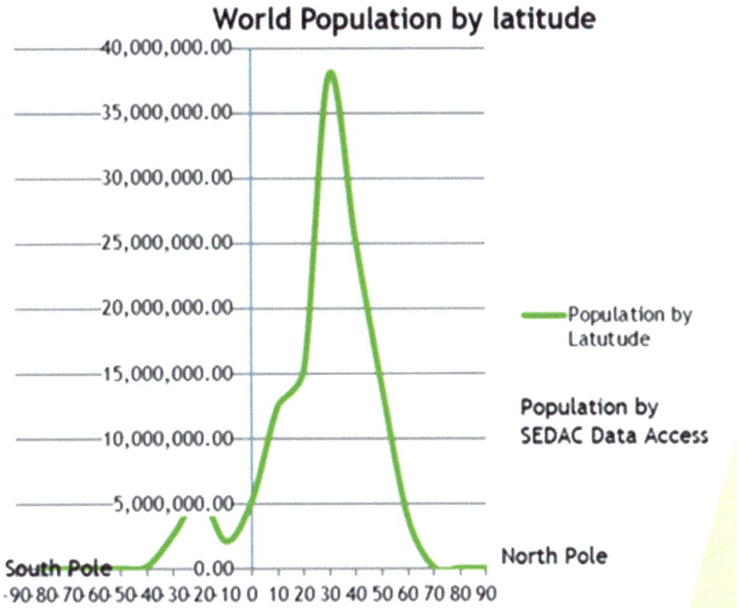

Figure 9. SEDAC Data Access Population (SEDAC 2020)

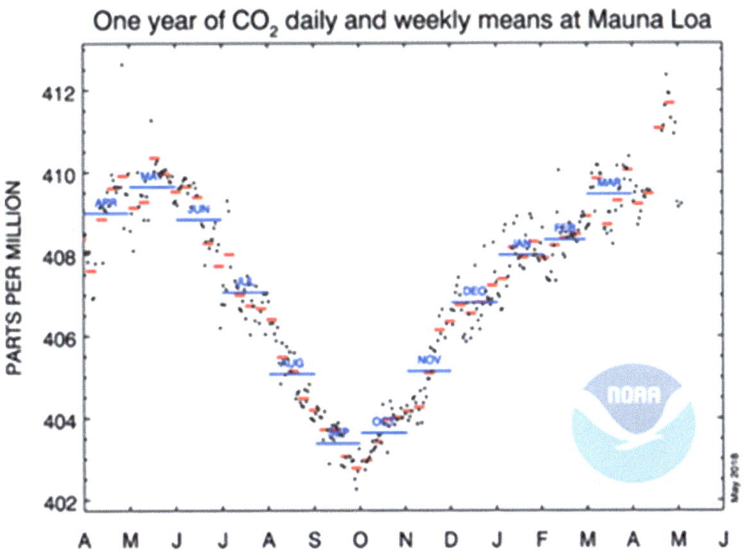

Figure 10. Mauna Loa oscillation. (NOAA weekly 2020)

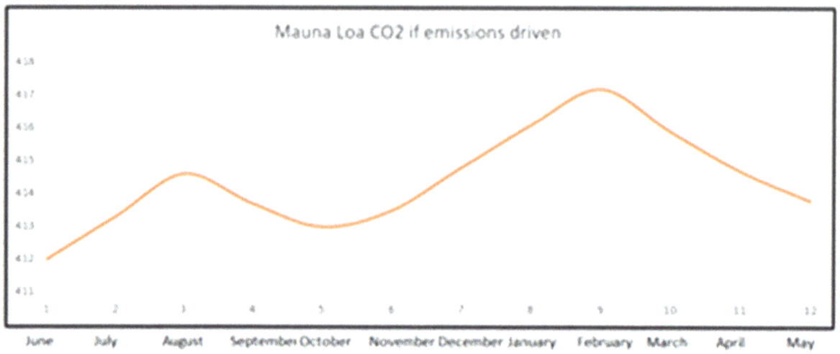

Graph 8. Mauna Loa if emissions driven (calculated) (D. White 2020)

(Q3) Are CO_2 cap and trade policies the best immediate intervention, or does globally increasing photosynthesis offer a more rapid and better long-term solution to climate change?

Atmospheric CO_2 Residence Time

A 2003 IPCC report states that atmospheric carbon dioxide residence time was between 5 to 200 years (IPCC WG I 2003). However, since 2003, the residence time has been increasing. This means we have to wait more than 150 years for a change in our carbon dioxide emissions to take effect. This is also, why atmospheric CO_2 is increasing even though the worldwide emissions of CO_2 have leveled off. Residence time of a system increases in either of two cases:

1. Input is greater than the drain can handle; 2. The drain is restricted. For atmospheric carbon

2. dioxide, the drain restriction is the issue. This is why residence time is increasing. Cap and Trade will never lower atmospheric CO_2. We need to increase photosynthesis consumption, the drain, which will lower atmospheric CO_2 to a value around 330 ppm by 2031. This cannot be accomplished by emissions work. Anything we do with CO_2 emissions have not and will not have any effect on atmospheric CO_2 for hundreds of years.

Below is a copy of the data collected on residence time (Table 2) based on the summary of Schwartz et al. of fifteen other manuscripts [35), and the constraints used (Graph 9). Even at a residence time of 100 years, Mauna Loa CO_2 never stays low. This is because we have a massive loss of photosynthesis consumption.

Facts: Residence time was once five years; now it is more than 200 years. If anyone were to take a survey at a climate change conference and ask the question, "What is the current residence time of atmospheric CO_2?" most scientists would say somewhere between 200 and 400 years. In my presentation at a conference, I said it was 500 years and no scientist questioned it. Another way to look at residence time is a signature from past events, which lowered CO_2 emissions. For example the oil embargo in the 1970's, multiple recessions and the big worldwide recession in 2009. The current COVID-19 pandemic. These are examples of lowered worldwide emissions.

Below is the current graph of Mauna Loa CO2. There is clearly no signature from these events.

Residence Time (Years)	Author	Year
700	Allen	2009
610	Zickfeld	2013
500	Matthews	2008
300	Plattner	2008
270	Cao	2010
230	Zickfeld	2012
220	Solomon	2012
220	Knutti	2012
210	Gillett	2011
180	Frolicher	2010
150	Hare	2006

Table 2. Published manuscripts stating atmospheric CO_2 residence time (Schwartz, 2018).

Employ the scientific method. First hypothesis. Then design of experiment to prove or disprove hypothesis. Then when enough data is collected to prove the hypothesis, the hypothesis changes to theory. If data doesn't prove hypothesis, then change the hypothesis to match what the data is saying. Never change the data to match the hypothesis. Then with sufficient data theory changes to scientific law.

We have an experiment on US 26 eastbound just west of Portland, Oregon. The hypothesis was the forested area on US 26E, by the Portland Oregon Zoo would consume most if not all the carbon dioxide from 160,000 vehicles per day. A permit was obtained from Oregon Department of Transportation to measure this. Sensors used are NIST certified and calibrated within one part per million. Graph 9 shows the rate of rise of atmospheric carbon dioxide less than 3 ppm/yr. The blue line represents the difference between the treed area and a non-treed area. Each location has a wind directional measurement. This measurement can confirm bad data from crosswind for example.

This experiment proves we can plant native shrubs and trees by roads and freeways instead of grass. The hypothesis is confirmed. This freeway has 161,000 autos per day on it, and approximately 460 auto exit (Sylvan exit 71) per day between the two sensor locations. The final day of testing was 6/12/2021. In September 2021the Scientific Law component of the experiment began. The newly obtained data confirmed last year's data at 16.8 ppm above the red line. We fully expect Scientific Law to be confirmed by June 2022.

Procedure: Place sensors at 6am daily for two weeks every other month for one year.

Pick up sensors at 7pm and analyze the data.

Put SD memory card from sensor into computer.

Import the data into an Excel spreadsheet named "data"

Repeat for the other sensor.

For each 10 seconds subtract the treed area from the non-tree area.

Sort data for "smallest to largest" from subtraction result.

Remove negative numbers in the subtraction result.

The negative numbers are from wind gusts. We tracked these many times.

Calculate average for the day.

Repeat.

There are several things to note in the graph. At no time did the blue line go below the red line. On December 20th, a very dark and rainy day the difference was 9 ppm. In April through June, we had very little rain. The graph shows this as lower difference. For photosynthesis we need these things, light, vegetation, moisture and carbon dioxide. Experiment Summary: This experiment proves we can plant native trees and shrubs instead of grass and they will eventually consume all the carbon dioxide from vehicles. This is applicable for ±50° from the equator.

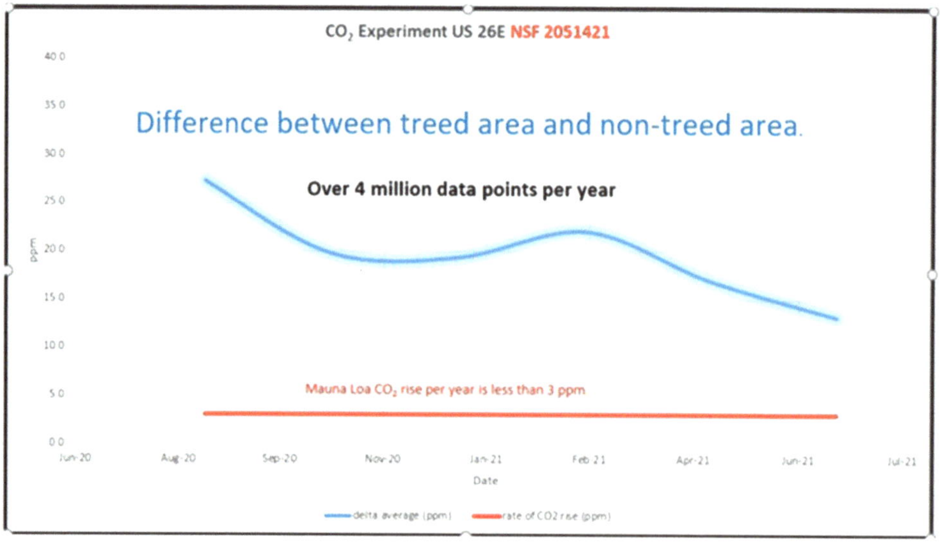

Graph 9. (D. White 2020-2021)

Atmospheric CO_2 is "extra," meaning that it is not consumed by photosynthesis.

Here are some important Assumptions and Constraints for Graph 10 below:

45% reduction in fossil fuel CO_2 emissions by 2030;
55% reduction in fossil fuel CO_2 emissions by 2130 due to depletion of those fuels;

By 2030, 45% reduction in the rate of rise of atmospheric CO_2.
By 2130, 45% reduction in CO_2 concentration
2230 55% reduction in CO_2 concentration and rate.

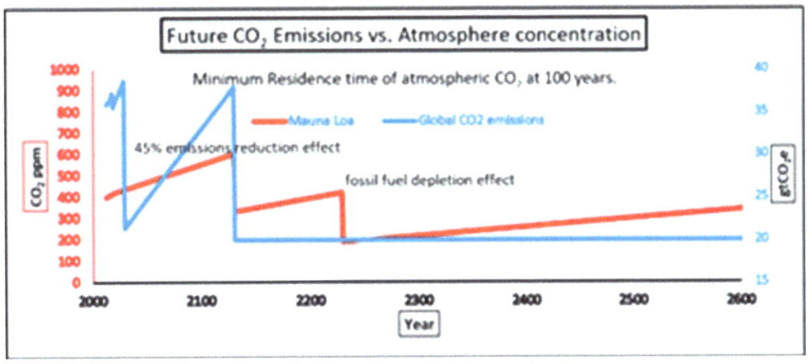

Graph 10. (D. White 2020)

Working to lower emissions will never lower atmospheric CO_2.

Even at a minimum residence time of 100 years, Mauna Loa carbon dioxide never stays low. We never reach equilibrium! (See Graph 9.)

Q4: What strategies can be employed to have the greatest positive impact over the upcoming crucial twelve-year period?

Planting trees is the single greatest weapon we have in the fight against climate change, and it is the only way to reduce atmospheric carbon dioxide to a value of 330 ppm by 2031 (Robert Scribbler 2020). When we plant trees *en masse* to regrow our forests, within twelve years we will be positioned to end climate change because this will increase the equilibrium from 8.6 GT to over 100 $GTyr^{-1}$ (See Graph 9).

Our survival and the promise of life for future generations is dependent, not upon onerous cap and trade policies or mitigation strategies, but rather upon planting trees. This increases the equilibrium from 8.6 $GTyr^{-1}$9.2 GT to over 100 $GTyr^{-1}$ within twelve years (See Graph 10). The world economic forum recently announced a resolution to plant 1 trillion native trees [28]. We heartily endorse this effort (World Economic Forum Jan 2020).

CONCLUSION

In this research paper, we have answered the four main questions in the abstract. We have shown two methods for determining the calculation of the correct equilibrium point (NetZero CO_2): first, by calculating photosynthesis worldwide; and second, by calculating that the decreasing oxygen level worldwide is below what is needed for life. We have also shown that atmospheric CO_2 is still rising even faster although the CO_2 emissions rise has slowed by 50%. Question 4 is answered here. We showed how cap and trade policies would have zero effect on the rise of atmospheric carbon dioxide because the equilibrium point is too low. The strategy with the most positive effect on lowering atmospheric CO_2 is by increasing photosynthesis.

This will in turn increase the equilibrium point to over 100 $GTyr^{-1}$. The only way to lower atmospheric carbon dioxide is to increase photosynthesis. The correct solution is to stop non-sustainable deforestation of large rainforests (such as those in India and the Amazon River Basin and its tributaries) and to plant 200 billion native trees and shrubs, especially in those areas that have been deforested. This will cause atmospheric CO_2 to lower to 330 ppm by 2031 (see Graph 11). Use ecosia.org for internet search engine. Ecosia.org plants trees.

Dr. Tom Crowther (Crowther et al. 2019) published a paper on increasing photosynthesis with recommendations of where to plant. The study found that most of the land suitable for restoring forests is in six countries: Russia (151 million hectares), USA (103 million hectares), Canada (78 million), Australia (58 million), Brazil (50 million), and China (40 million). Appendix 1 shows how to plant the trees and shrubs. This will increase consumption of CO_2 to over 100 Gt yr^{-1}. Not every forest hectare is equivalent in photosynthesis consumption of CO_2. Rainforests consume 90-100 tons per hectare of carbon dioxide per annum. Other forests are from one quarter to 8 tons per hectare per annum.

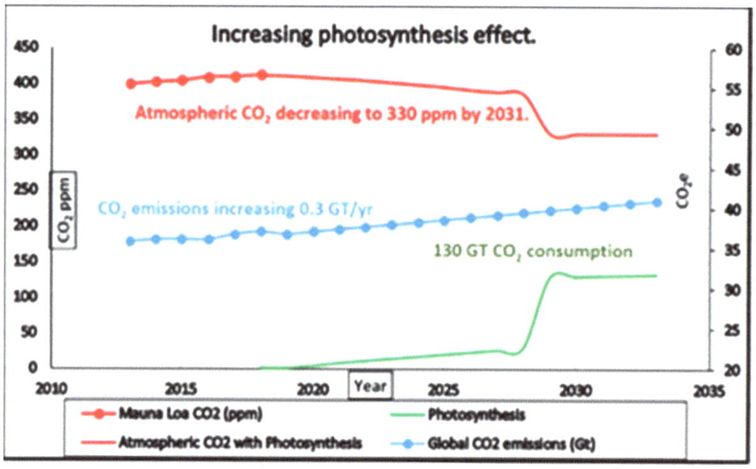

Graph 11. Increasing Photosynthesis Effect (D. White 2020)

Acknowledgments

This research article has received no funding from any outside sources. All work performed by the authors has been done without any compensation. The goal of Climate Change Truth, Inc. is to follow the data. The goal of The Energetic Health Institute is to share Information from the heart for the betterment of all.

Data Availability Statement

Data will be available in the link provided. https://www.ncei.noaa.gov/

Conflict of Interest Statement

The authors have no conflict of interest that we are aware of.
Appendix 1.

Ideas on How to Plant Trees and Shrubs

Following is a recommended government policy guide to lower atmospheric carbon dioxide quickly. We can keep emissions at 37 Gt yr^{-1} without any new reduction plans. A refocus on planting trees and shrubs is required. Native plants that produce oxygen year-round are preferred.

1. Put pressure on Brazil and other Amazon Rainforest countries to stop deforestation as soon as possible. Stop the biomass burning that puts 300 million tons of carbon dioxide into the atmosphere each year, which causes the switching of the rainforest to an oxygen sink and CO_2 producer. This switching caused the recent rise in atmospheric carbon dioxide concentration of 53 ppm (D. White, 2019). Then, after ten years, finish burning what is needed at 10% per year for ten years.

2. Provide space where the public can come and plant trees and shrubs on government -owned lands. The cost would be minimal. A website could be created to document each planting area.

3. Plant shrubs in all freeway medians and sides. This would pay for itself within two years because of lower maintenance costs. Plant native shrubs at a minimal spacing so all light is used in photosynthesis. This will take in 1 ton of CO_2 emissions per acre per year right at the source of auto emissions. The space would not need to be mowed every week in the summer.

4. Get schools involved to plant massive numbers of trees and shrubs on their property and on government property as in 1 above.

5. Add trees and shrubs to parks.

6. Give tax incentives for businesses to plant trees and shrubs. People can plant shrubs on roofs which can structurally handle dirt with minimal spacing and drip irrigation, creating "green roofs."

7. Attend to wildfires quickly. Get a retainer for a jet plane to use from the start of any wildfire.

These action steps worldwide, will increase carbon dioxide consumption by 2-3 billion tons per year (not including the effects of rainforest renewal, which eventually will consume 60-100 Gt yr^{-1}). All embassy environmental scientists have concurred with this science and have encouraged their countries to plant trees. China is planting millions of trees. India stopped deforestation of its rainforest and is planting trees . Pakistan has already planted 4 billion trees and will plant seven billion more in the next four years. Since May 2018, these countries have planted more than 4 billion trees.

References:

2017 data http://www.repositioncruises.com/cruise-industry/#statistics

A. Baccini Et al. 2017 Tropical forests are a net carbon source based on aboveground measurements of gain and loss. Science 13 Oct 2017: Vol. 358, Issue 6360, pp. 230-234 DOI: 10.1126/science.aam5962

Brienen, R. *et al.* 2015 Long-term decline of the Amazon carbon sink. *Nature* 519, 344–348 (2015). https://doi.org/10.1038/nature14283

Emily Chasan 2019 "We Already Have The World's Most Efficient Carbon Capture Technology"

Bloomberg Businessweek August 2019. https://www.bloomberg.com/news/features/2019-08 02/we-already-have-the-world-s-most-efficient-carbon-capture-technology

Crowther et al. 2019 *Science* 05 Jul 2019: Vol. 365, Issue 6448, pp. 76-79 DOI: 10.1126/science.aax0848

D. White, 2019 Discovery: Reduction in Photosynthesis Correlation to Carbon Dioxide Increase, ACTA SCIENTIFIC AGRICULTURE (ISSN: 2581-365X) Volume 3 Issue 4 April 2019 https://actascientific.com/ASAG/pdf/ASAG-03-0393.pdf

D. White 2020 Statistical regression https://issues.pangaea.de/secure/attachment/142070/regression.xlsx

Ethiopia 2019 Ethiopia Plants Over 350 Million Tress In A Day, Setting New World Record" United Nations Environment August 2019. https://www.unenvironment.org/news-and-stories/story/ethiopia-plants-over-350-million-trees-day-setting-new-world-record

Ficks diffusion 2018 http://cctruth.org/Ficks%20diffusion.pdf

Flight Center, Greenbelt, 5 Maryland, USA https://sites.agu.org/newsroom/files/2015/09/2015GB005139.pdf

Natalya Gallo 2014 Ocean Deoxygenation http://oceanscientists.org/index.php/topics/ocean-deoxygenation

Global Carbon Atlas 2020 http://www.globalcarbonatlas.org/en/CO2-emissions

CHRISTINE L. GOODALE et al. 2002 FOREST CARBON SINKS IN THE NORTHERN HEMISPHERE,

Ecological Applications, 12(3), 2002, pp. 891–899 q 2002 by the Ecological Society of America
http://www.eeb.cornell.edu/goodale/2002%20GoodaleEcolAppl.pdf

Jianping Huang et al 2018 The global oxygen budget and its future projection
https://www.sciencedirect.com/science/article/pii/S209592731830375X

IPCC WG I 2003 https://archive.ipcc.ch/ipccreports/tar/wg1/016.htm

E. Kintisch, 2017 Amazon rainforest ability to soak up carbon dioxide is falling, doi:10.1126/science.aab0336, http://www.sciencemag.org/news/2015/03/amazon-rainforest-ability-soak-carbon-dioxide-falling, 2017

Machida et aL. 2002 VERTICAL AND MERIDIONAL DISTRIBUTIONS OF CO2
https://ps.uci.edu/~rowlandblake/publications/177.pdf

Kenzie Mastroe 2019 https://www.sheknows.com/living/articles/2086062/amazon-rainforest-fires-2019/
NOAA gr 2020 https://www.esrl.noaa.gov/gmd/ccgg/trends/gr.html

NOAA weekly 2020 https://www.esrl.noaa.gov/gmd/ccgg/trends/weekly.html

Oxygen Levels 2019 https://www.oxygenlevels.org/

Rogelj, J et al. 2018: Mitigation Pathways Compatible with 1.5°C in the Context of Sustainable Development. In: Global Warming of 1.5°C. An IPCC Special Report on the impacts of global warming of 1.5°C above pre-industrial levels and related global greenhouse gas emission pathways, in the context of strengthening the global response to the threat of climate change, sustainable development, and efforts to eradicate carbon dioxide.

C. Rose, et al 2015, The Characterization of Feces and Urine: A Review of the Literature to Inform Advanced Treatment Technology, Crit Rev Environ Sci Technol. 2015 Sep 2; 45(17): 1827–1879.
Max Roser 2015 Very Long-Term Perspective on Deforestation in Specific Regions
https://ourworldindata.org/forests

Cecile S. Rousseaux et al 2015 Global Modeling and Assimilation Office, NASA Goddard Space

SEDAC 2020 Center for International Earth Science Information Network - CIESIN - Columbia University. 2018. Population Estimation Service, Version 3 (PES-v3). Palisades, NY: NASA Socioeconomic Data and Applications Center (SEDAC). https://doi.org/10.7927/H4DR2SK5

Hans R. Schneider et al. 1989 An Evaluation of the Role of Eddy Diffusion in Stratospheric Interactive Two-Dimensional Models https://doi.org/10.1175/1520-0469(1989)046<2079:AEOTRO>2.0.CO;2

Schwartz, 2018 Unrealized Global Temperature Increase: Implications of Current Uncertainties, 07 March 2018, DOI: (10.1002/2017JD028121)

Robert Scribbler 2020 https://robertscribbler.com/2016/08/05/carbon-sinks-in-crisis-it-looks-like-the-worlds-largest-rainforest-is-starting-to-bleed-greenhouse-gasses/

Takahashi 1997 Ocean Surface CO2 https://www.pmel.noaa.gov/co2/story/LDEO+Surface+Ocean+CO2+Climatology

Takahashi, Taro et al. 2008 Climatological mean and decadal change in surface ocean pCO2, and net sea–air CO_2 flux over the global oceans. https://doi.org/10.1016/j.dsr2.2008.12.009

Welty et al, 1984 Fundamentals of Momentum Heat and Mass transfer. http://bcs.wiley.com/he-bcs/Books?action=index&bcsId=4049&itemId=0470128682

World Economic Forum Jan 2020. https://www.weforum.org/agenda/2020/01/one-trillion-trees-world-economic-forum-launches-plan-to-help-nature-and-the-climate/

PMEL ocean acidity data. https://www.pmel.noaa.gov/co2/story/OA+Observations+and+Data

Comments from Scientific Peer Review

First Atmospheric Carbon Dioxide equilibrium manuscript to define NetZeroCO2E is peer reviewed and published in worlds top climate change journal. The Journal of Earth Science & Climatic Change is the top most climate change Journal by impact factor! This manuscript received the following comments from the review.

1. First of all I congratulate author for an amazing article about "The Essential Role of Photosynthesis in Defining Net Zero Carbon Dioxide Emissions for Equilibrium Calculations."

2. It is well written and well explained about all the points.

3. In this study author explained about current climate change conversation and answered four main questions, by calculating photosynthesis, decreasing oxygen levels worldwide and also explained about the atmospheric Carbon dioxide is still rising even faster although the Carbon dioxide emissions rise has slowed by 50%.

4. He explains that how cap and trade policies would have zero effect on the rise of atmospheric carbon dioxide because the equilibrium point is too low. The strategy with the most positive effect on lowering atmospheric CO2 is by increasing photosynthesis.

5. Nicely done research work. Clearly explained with tables, figures and graphs.

6. Author explained how to increase the equilibrium point to over 100 GTyr-1. By only way to lower atmospheric carbon dioxide is to increase photosynthesis. The correct solution is to stop non-sustainable

deforestation of large rain forests (such as those in India and the Amazon River Basin and its tributaries) and to plant 200 billion native trees and shrubs, especially in those areas that have been deforested.

7. Author explained about planting trees to have the greatest positive impact over the upcoming crucial period.

8. There are many positive points which is useful for everyone to understand and learn from it, the article is very impressive.

Key Findings:

The Northern Hemisphere forests only consume 2.6 billion tons of carbon dioxide per year through photosynthesis.

All the southern hemisphere forests have switched to be a oxygen sink and carbon dioxide producer due to organic decay.

Ocean photosynthesis is decreasing.

The tropospheric carbon dioxide is diffusing to the exosphere not the ocean. The ocean is not a sink for carbon dioxide.

Experiment which proves we can plant native trees and shrubs near roads (where applicable) and in 10 years they will consume all the carbon dioxide from the vehicles.

Chapter 2 Questions:

1. What is the total photosynthesis worldwide in 2022?
2. How much photosynthesis is in the Northern Hemisphere forests?
3. Is ocean photosynthesis declining?
4. Are the oceans a sink for tropospheric carbon dioxide?
5. What is the average residence time for tropospheric carbon dioxide?
6. At what latitude do most people live in the world?
7. Describe the scientific method.

Chapter 3

Green House Gases

A few years ago, we contacted the National Academy of Sciences and spoke with Dr. Mike Kuperberg, Executive Director of the U.S. Global Change Research Program (USGCRP). He recognized the integrity of the research and sent my analyses to the other scientists in their office. Their consensus was to have me assemble a team to participate in the annual “Expert and Government Review (EAGR)” program of the Intergovernmental Panel on Climate Change (IPCC) Reports.

A team of PhD’s was assembled, whose ranks soon swelled to thirty other scientists who are also contributors to this textbook. Together we participated in the “EAGR” program, and unanimously identified all manner of inaccurate and incompetent science in their reports. Furthermore, we were instrumental in the firing of Adam Yeeley, the chief editor of *Nature Climate Change,* whose doctorate was in political science. He had been permitting IPCC scientists to publish loosely referenced manuscripts and circular reference them in their reports. Needless to say, this is not genuine science.

In fact, the IPCC reports might even be described as deliberate science fiction. The writers identify themselves as climate experts and inform governments globally in their reports on what to believe about climate change. These fallacious reports lead to misguided government policies that negatively impact every person and business around the globe through unnecessary economic restrictions and taxation.

Our thirty-PhD review of IPCC working Group 1, first order draft for Ar6, exposed their faulty global warming potential model. This model assumes equal greenhouse gas (GHG) concentrations, which will never occur in nature. For example, carbon dioxide is more than 200 times the concentration of methane. Furthermore, in Group 1, we found in Annex 2, a table with the correct order of GHG effects. Any model which ignores data to benchmark it with this correct order is a fake. We sent our review at least twenty-three times to inform them they had to benchmark their Annex 2 table to the correct order of GHG effects. However, for the final draft for Ar6 they chose not to benchmark their final draft but instead chose to delete the table in Annex 2, which still left their fake GWP model intact. This wasn’t just overlooking the benchmarking of the data. They intentionally hid the fact that their science model was false. This is a blatant example of cooking the data to mislead policymakers.

Carbon dioxide is not pollution! Everyone breathes it out! The recent Supreme Court ruling very much strengthens our legal argument for the lawsuits on cctruth.org lawsuit page to stop cap and trade! Exhibits 1 and 2 are our thirty-PhD reviews of the Intergovernmental Panel on Climate Change (IPCC) reports, which are deliberate science fiction.

What the Supreme Court did to the corrupt Environment Protection Agency (EPA):

The Clean Air Act of 1967 directed the EPA to tackle such issues as Acid Rain and others. It instructs the EPA to make a "toxic chemicals" list. Then any chemical the EPA wants to regulate in Section 111 subsection d must be on the toxic chemicals list. The EPA in 2015 illegally used section 111 subsection d to go after greenhouse gases without putting them on the toxic chemicals list. Carbon dioxide, methane etc. are not toxic and therefore could not be put on the list. It would behoove Congress to open a congressional investigation and put the people who did this on trial. If congress wants the EPA to regulate Greenhouse gases, then it needs to write another law. However, they need to keep the following scientific facts in mind when writing such legislation.

Greenhouse Gas	Emission band cm^{-1}	Radiation Flux W/m^2
H_2O, water vapor	All bands	303.8
CO_2, carbon dioxide	All bands	30.9-37.3
CH_4, methane	1200-1400	1.0-1.2
N_2O, nitrous oxide	1200-1300	1.1-1.3
O_3, ozone	900-1100	3.0-3.3

Table 3.1 Georgia Tech data f

http://irina.eas.gatech.edu/EAS8803_SPRING2012/Lec7.pdf

Calculating percent effect is for water vapor 303.8/344.2*100=88.26%

The 344.2 is the sum of these effects.

Water vapor is 88.26 ghg effect.
Carbon dioxide is 10.17% ghg effect.
Ozone is 8.92% ghg effect.
Nitrous Oxide is 0.334% ghg effect.
Methane is 0.33% ghg effect.

Learn more about Greenhouse Gases here: Greenhouse Gases - Climate Change Truth (cctruth.org)

Dr. T. J. Blasing of Oak Ridge National Laboratory exposed greenhouse gases to long wave radiation (2-2). Table 2.2 has his results.

Gas (Watts/m2)	Increased radiative forcing
CO_2 (carbon dioxide) ppm.	1.94
CH_4 (methane) ppb.	0.50
O_3 (ozone)	0.40
N_2O (nitrous oxide) ppb	0.20

The remainder are negligible.

Chapter 3 Questions:

1. Why is the IPCC global warming potential model fake?
2. Why do the Georgia Tech and Dr. Blasing's data differ? How are they the same?

References:

Current Greenhouse Gas Concentrations (lbl.gov)

Table 7.2 page 3 of http://irina.eas.gatech.edu/EAS8803_SPRING2012/Lec7.pdf

Chapter 4

Astrophysical Warming

The following graphs of Global Temperature are produced with NOAA station data and NOAA Software GHCNM-v3.3.0 running on our server. Land surface temperatures are available from the Global Historical Climate Network-Monthly (GHCN-M). These will be automatically updated the 9th of each month and available on the 10th of the month. GHCNM (version 3): J. H. Lawrimore, M. J. Menne, B. E. Gleason, C. N. Williams, D. B. Wuertz, R. S. Vose, and J. Rennie (2011), An overview of the Global Historical Climatology Network monthly mean temperature data set, version 3, J. Geophys. Res., 116, D19121, doi:10.1029/2011JD016187.

The graphs below are created by our Unix admin with help from NOAA Dr. Jay Lawrimore. The first graph is of all hemisphere stations and shows warming of about 1.1 C since 1975. The second graph is the 80° northern hemisphere and shows 9.9 degrees warming since 1975. The third is the northern hemisphere stations and shows about a 1.1 C increase since 1975. The fourth is the southern hemisphere stations and shows about a 0.8 C increase since 1980. The fourth graph is -50 degrees below the equator to the south pole. This shows no warming.

Therefore, the Antarctic glaciers are not receding like the Arctic ones. Greenhouse gases, like all gases, diffuse until they are equi-distant apart; thus, they cannot cause this type of warming. Any freshman college chemistry student knows this. Also, these and all gases get distributed throughout the earth by winds. We hope to work soon with Oregon State University Chemical, Biological and Environmental Engineering Professor of Transport II HEAT CHE 332 to determine how much of this heat is caused by GHG.

It's no wonder a large percentage of scientists on the climate deniers' list online are Astrophysics PhD's. **The shift of the earth's tilt in the year 2000-2004 has the following effect of temperature rise by latitude. You can see this effect in individual graphs on the Graphs page. This shift is described by NASA JPL scientists as massive.**

80° North effect is 4.6 C
70° North effect is 3.7 C
60° and 50° North show a lesser effect.

The first astrophysical event was in 1975. This effect is 1.0 C rise at the north pole.

You can see all the NOAA Graphs here. [NOAA Data Station Graphs by Latitude]

NOAA National Centers for Environmental information, Climate at a Glance: Global Time Series, published March 2018, retrieved on April 7, 2018 from http://www.ncdc.noaa.gov/cag/ However you can't choose all longitudes currently. You can see the mapping has warming in the north and cooling in the south.

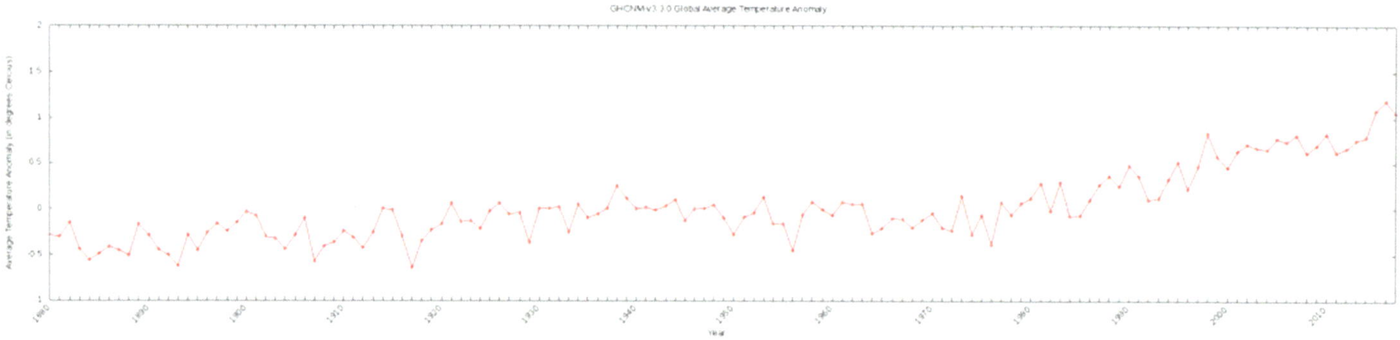

Figure 4-1 Global Average Temperature

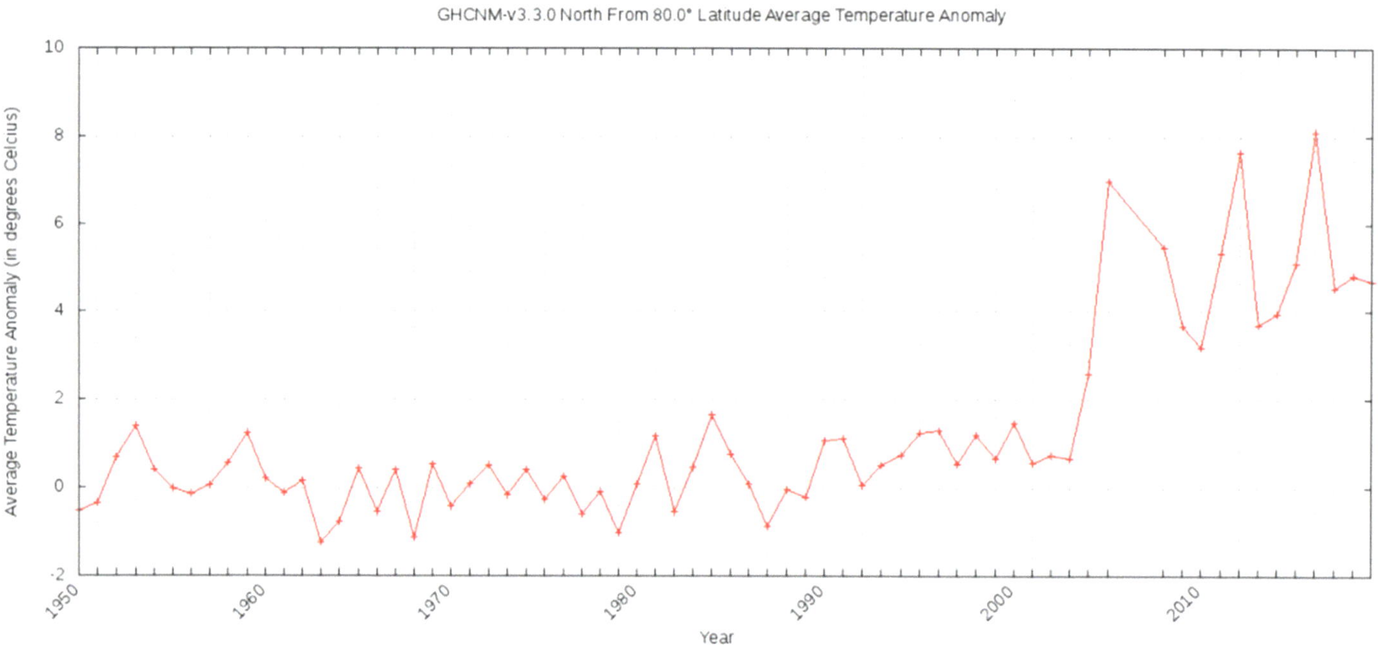

Figure 4-2 North 80 degrees' temperature graph

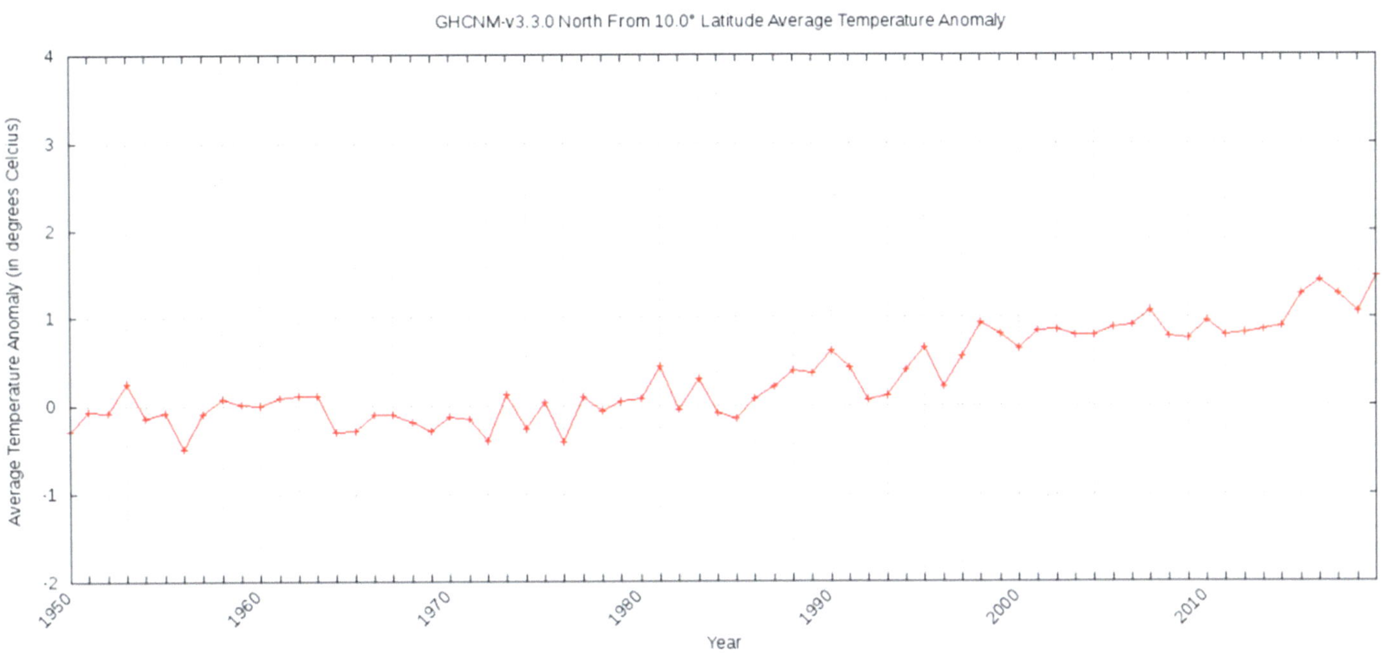

Figure 4-3 North 10 degrees (just above equator)

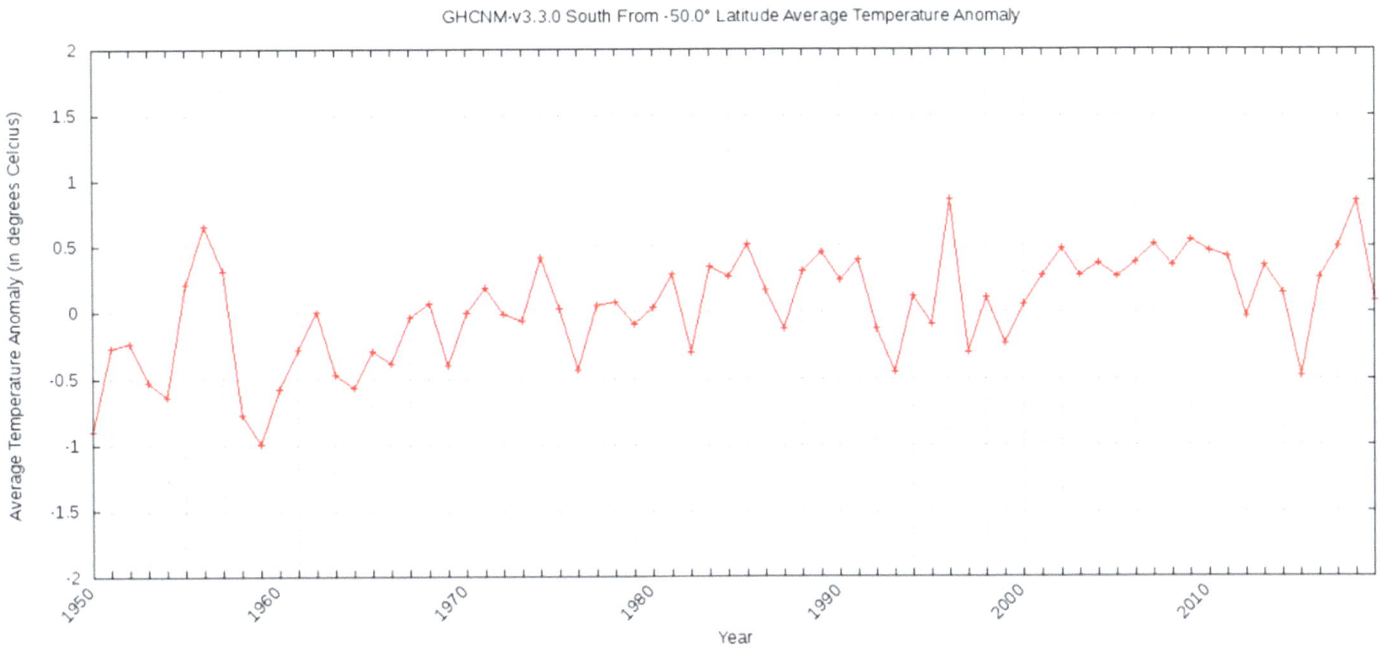

Figure 4-4 South 50 degrees. Almost no warming,

The script we made to create the graphs has been confirmed by NOAA to be correct. The link below will allow anyone with access to a Unix server to install the NOAA software and create these graphs: Graph Script

The graph below is from NOAA data station temperature change by latitude with January 2024 data and before.

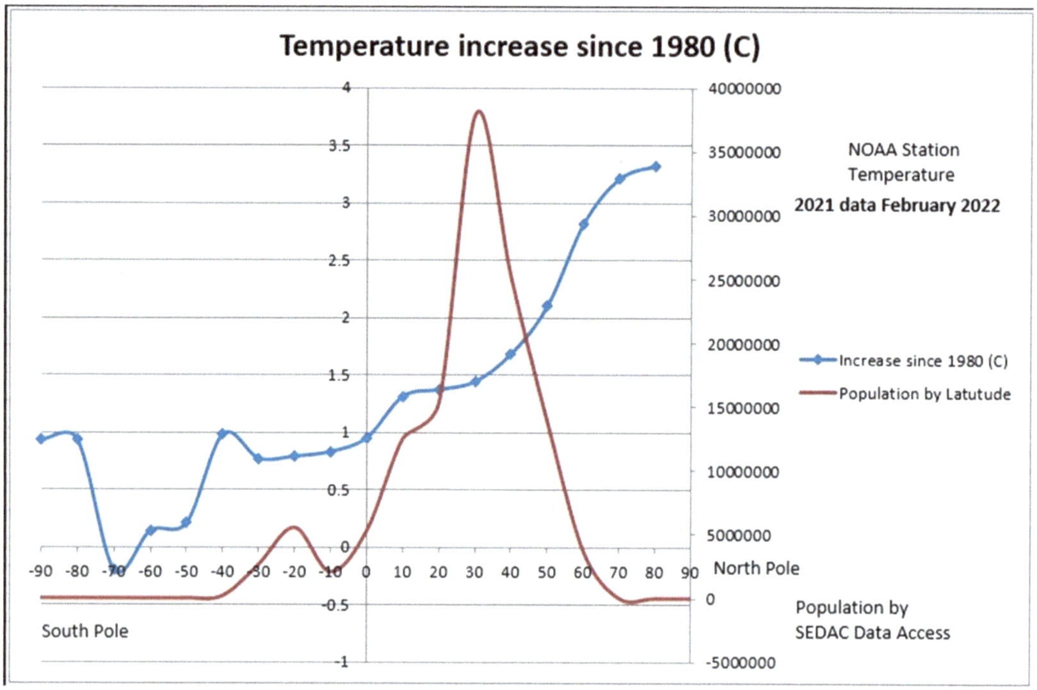
Temperature increase since 1980 (C)
NOAA Station
Temperature
2021 data February 2022
Increase since 1980 (C)
Population by Latutude
North Pole
South Pole
Population by
SEDAC Data Access

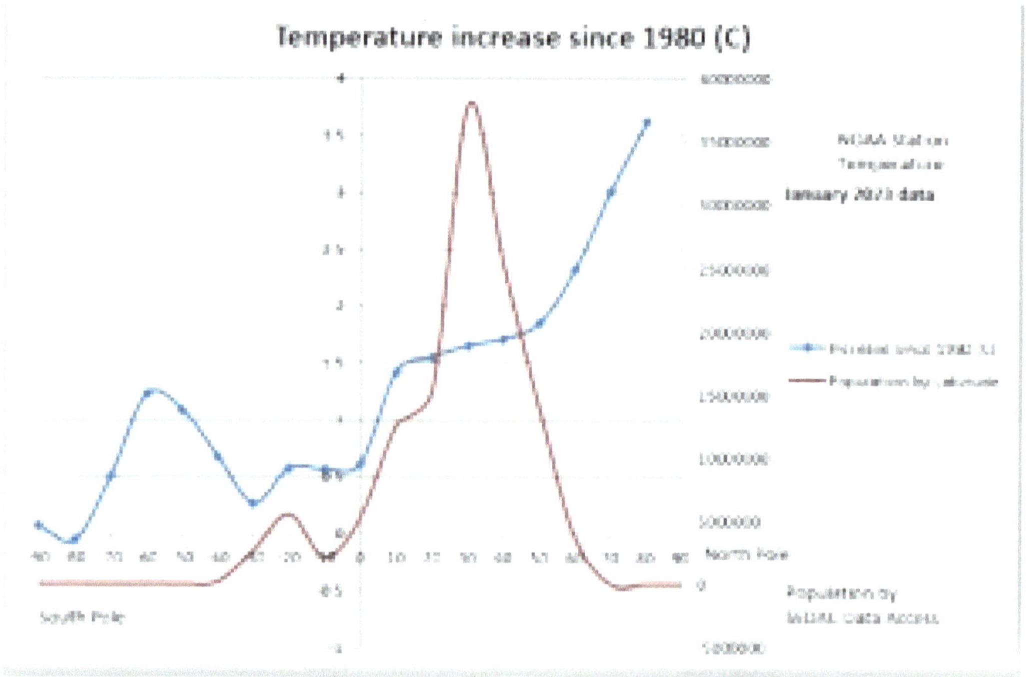
Temperature increase since 1980 (C)

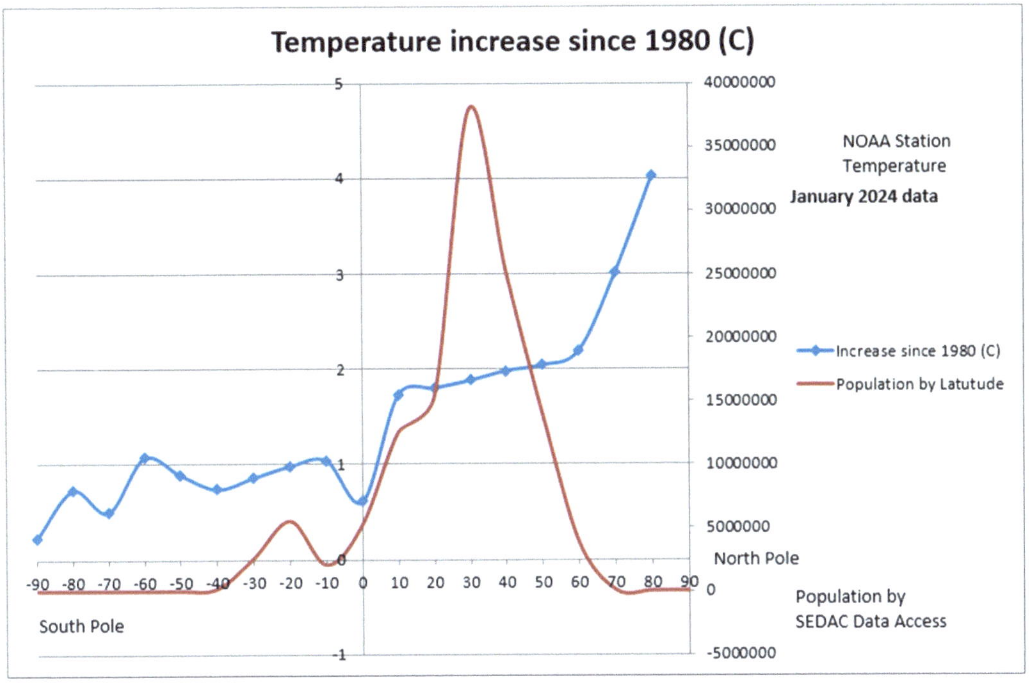

Figure 4-5 Minimum temperature increase is 0.23 C. Max is 4.04 C

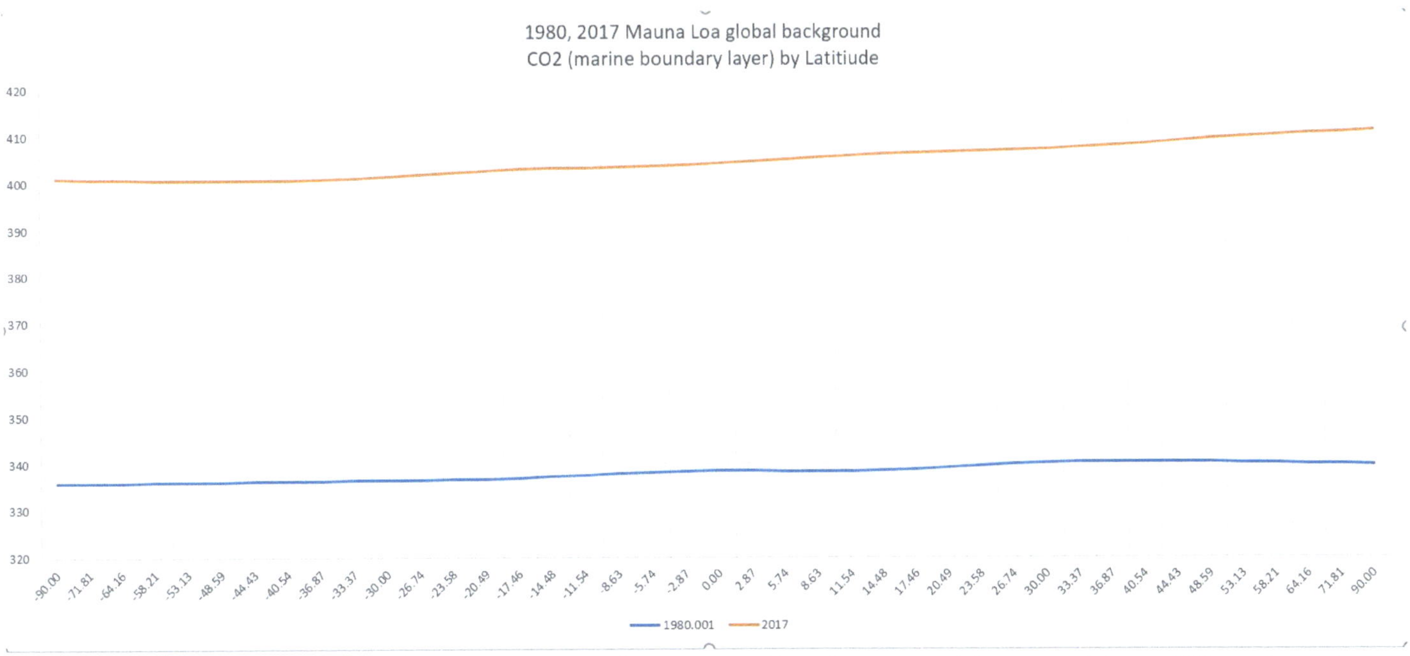

Figure 4-6 Carbon dioxide minimum 401 ppm. Maximum 410 ppm. Very little signature to temperature. The rise of carbon dioxide by latitude is only 2.5%. This data is courtesy of Dr. Pieter Tans of NOAA and Dave White of cctruth.org.

On June 28th 2023 the News Media reported that the tilt of the earth had shifted. This lends more credence to the astrophysical (earth tilt) causing the warming we experience. https://www.smithsonianmag.com/smart-news/humans-have-shifted-earths-axis-by-pumping-lots-of-groundwater-180982403/

Furthermore, in 2014 a NASA JPL manuscript published in the farce journal Nature Climate Change said a massive shift in the tilt happened in 2003 caused by melting Greenland Glaciers. They swapped cause and effect.

Chapter 4 Questions:

1. Web search and find a shift in the tilt of the earth in the 1970's.
2. Web search and find a shift in the tilt of the earth around 2002-2003.
3. Why do 90% of earth's inhabitants feel like there is global warming?

Chapter 5

Residence Time of Atmospheric Carbon Dioxide

Residence time is like standing water in a kitchen sink with the drain plugged. The water is residing longer. Retention time is the same idea as residence time. There are two types of residence time:

1. Average residence time.
2. Instantaneous residence time.

The average residence time is the average time that a molecule of carbon dioxide, for example, stays in the troposphere.

The Instantaneous residence time is the time for equilibrium to be achieved after some disturbance. For example, a large injection of carbon dioxide into the troposphere would raise the concentration of carbon dioxide overall. However, it would take some days for the atmospheric winds to even out the disturbance.

Average Residence Time of Atmospheric CO_2

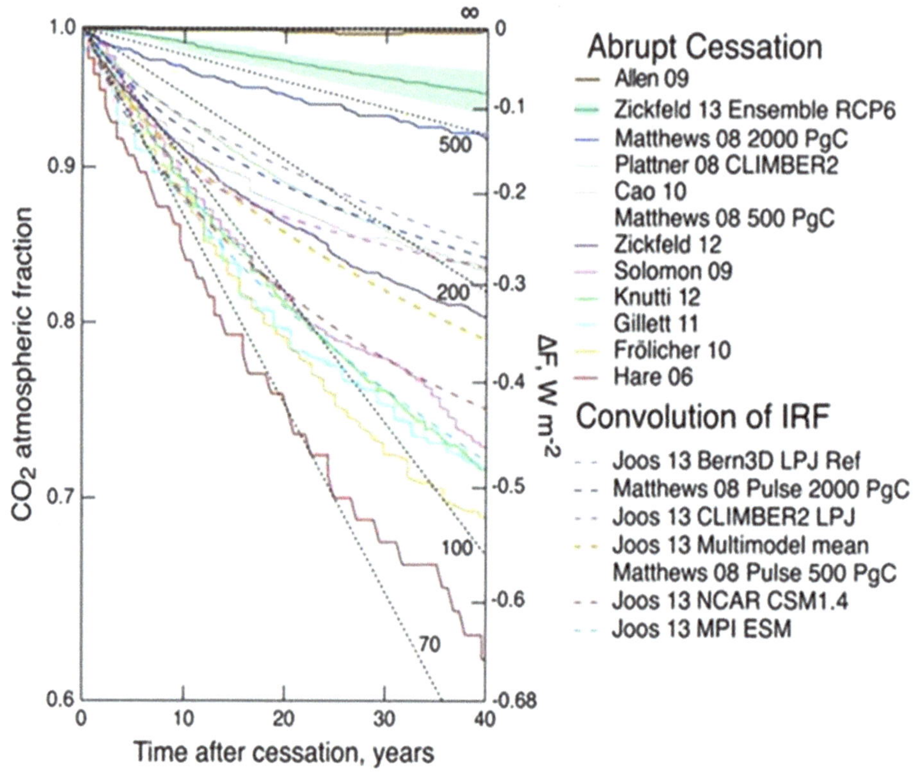

Residence Time (Years)	Author	Year
>700	Allen	2009
610	Zickfeld	2013
500	Matthews	2008
300	Plattner	2008
270	Cao	2010
230	Zickfeld	2012
220	Solomon	2012
220	Knutti	2012
210	Gillett	2011
180	Frolicher	2010
150	Hare	2006

Another way to look at residence time is a signature from past events, which lowered CO_2 emissions:

- Oil embargo in the 1970's;
- Multiple recessions;
- Worldwide recession in 2009;

•COVID-19 pandemic.

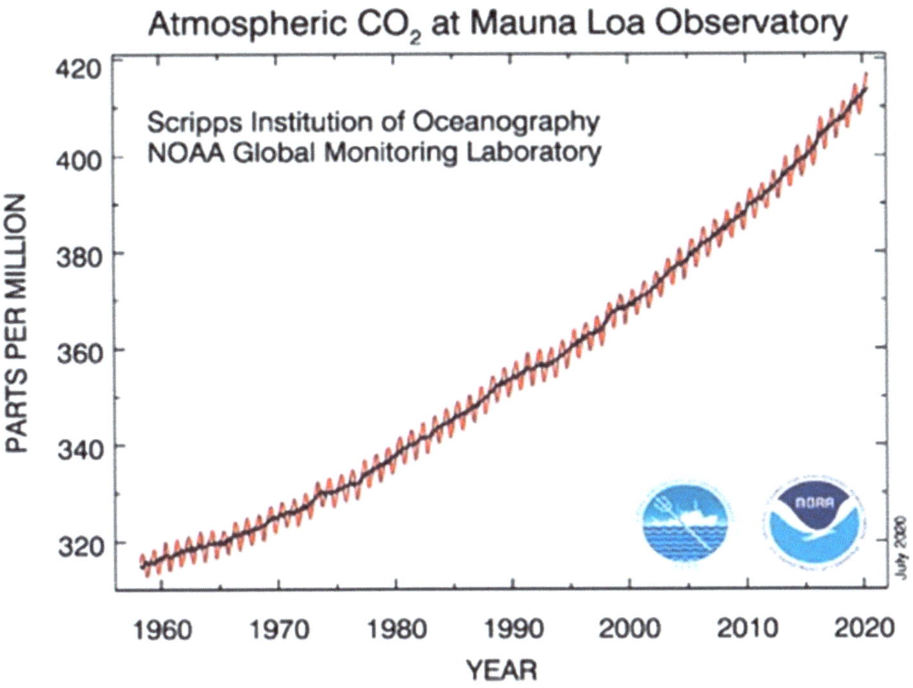

It is clear that there is no signature from these events.

Chapter 5 Questions:

1. What is the difference between average residence time and instantaneous residence time?
2. How long on average will it take to remove a molecule by Photosynthesis from the troposphere?
3. Why is the residence time increasing? (Check chapter 1)

Reference:

Unrealized Global Temperature Increase: Implications of Current Uncertainties, Schwartz, S. E. J. Geophys. Res. , 2018, doi: 10.1002/2017JD028121.

Chapter 6

NOAA Mauna Loa Data and Fraud

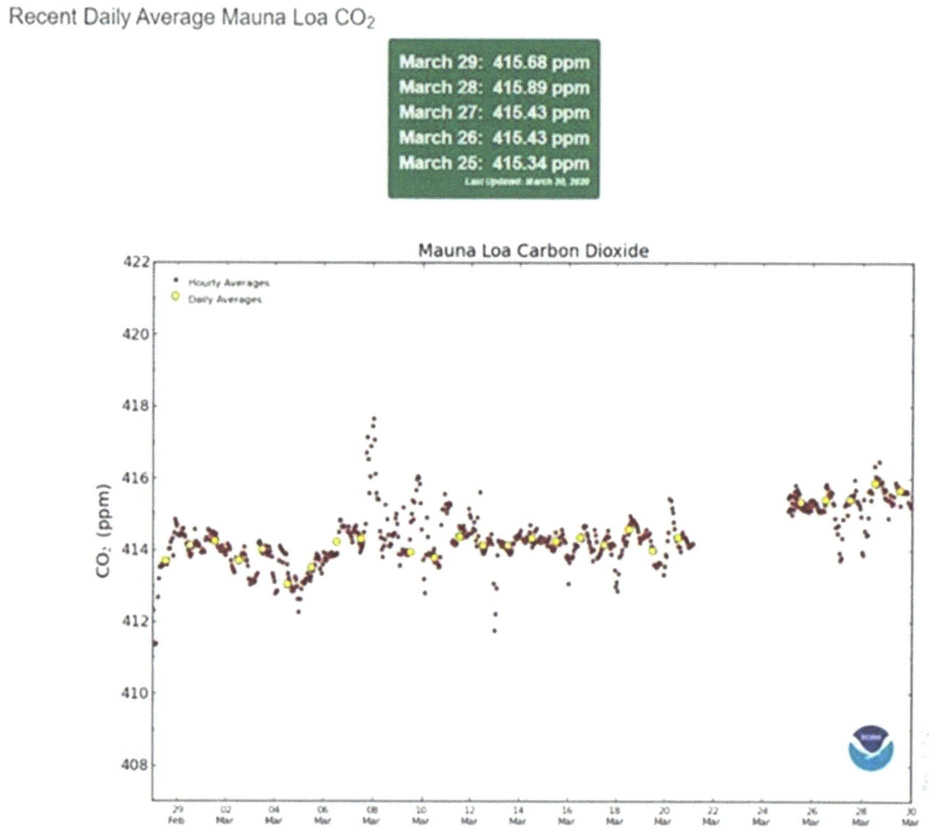

Graph 6-1 Flat carbon dioxide from 20 billion trees planted.

After the analyzer was “fixed,” NOAA Mauna Loa CO_2 data Increased by exactly 1.500 ppm.

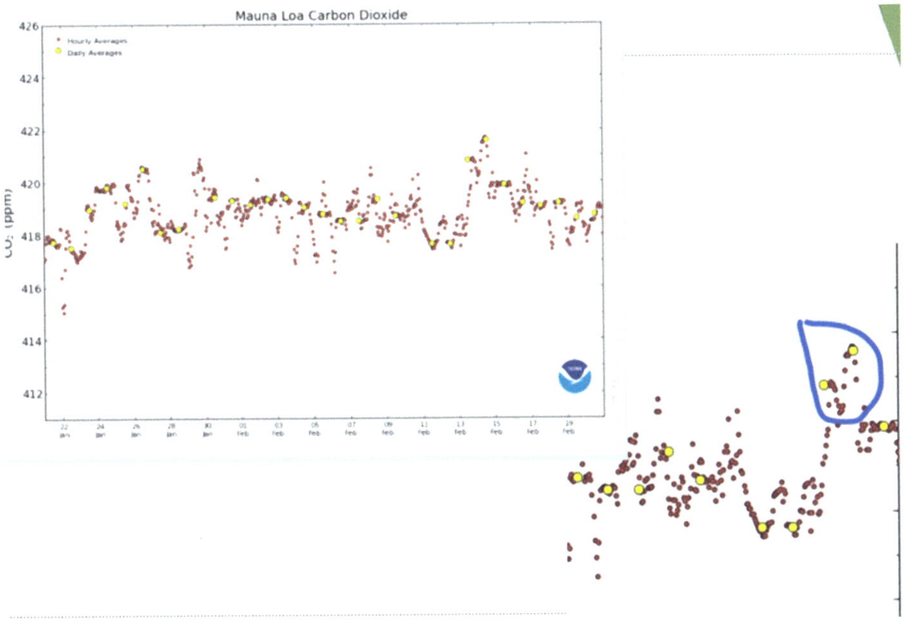

Graph 6-2 Direct increase of 3 ppm over two days. This never had happened previously.

After twenty-five billion trees had been planted, we had a flat carbon dioxide rise during the January time frame. For seventy years, January timeframes had increased by 1.5 ppm per month. This year: no rise. NOAA didn't like it, so they committed fraud.

For seventy years Mauna Loa worldwide carbon dioxide peaked in late May and decreased until early November. However, Graph 6-3 shows a peak this year in February despite a stoppage of deforestation in India and Peru and twenty-eight billion trees planted since May 2018. The red enclosed area is fraud!

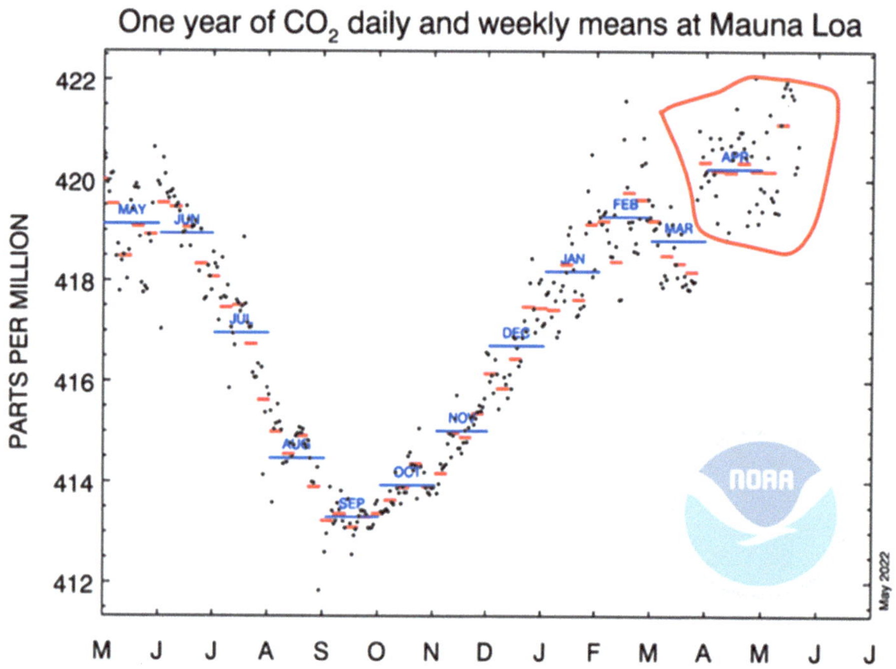

Graph 6-3

Recent Daily Average Mauna Loa CO_2

September 23: 414.81 ppm
September 22: 414.73 ppm
September 21: 416.31 ppm
September 20: 416.32 ppm
September 19: 416.19 ppm
Last Updated: September 24, 2022

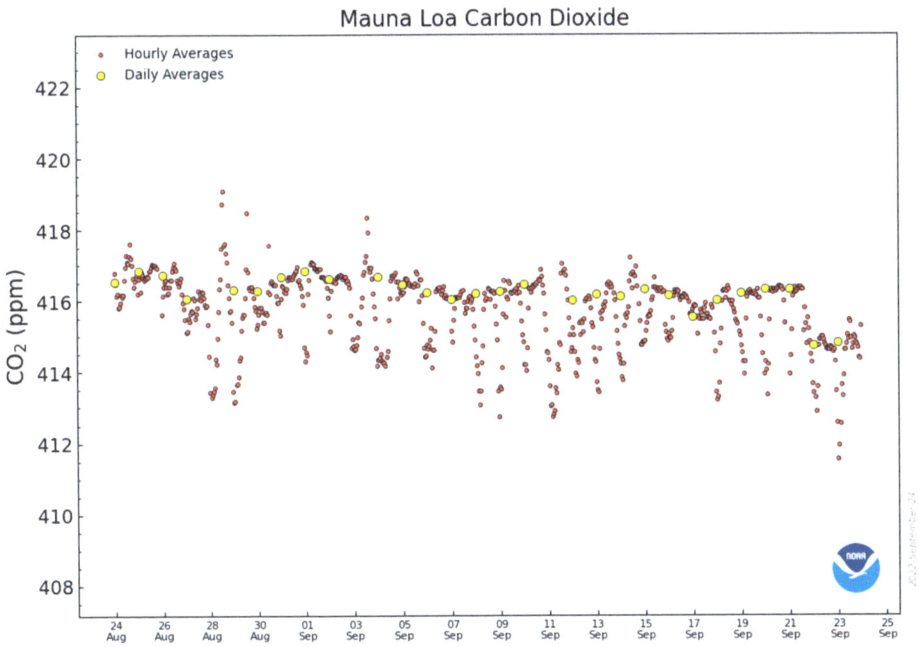

Graph 6-4

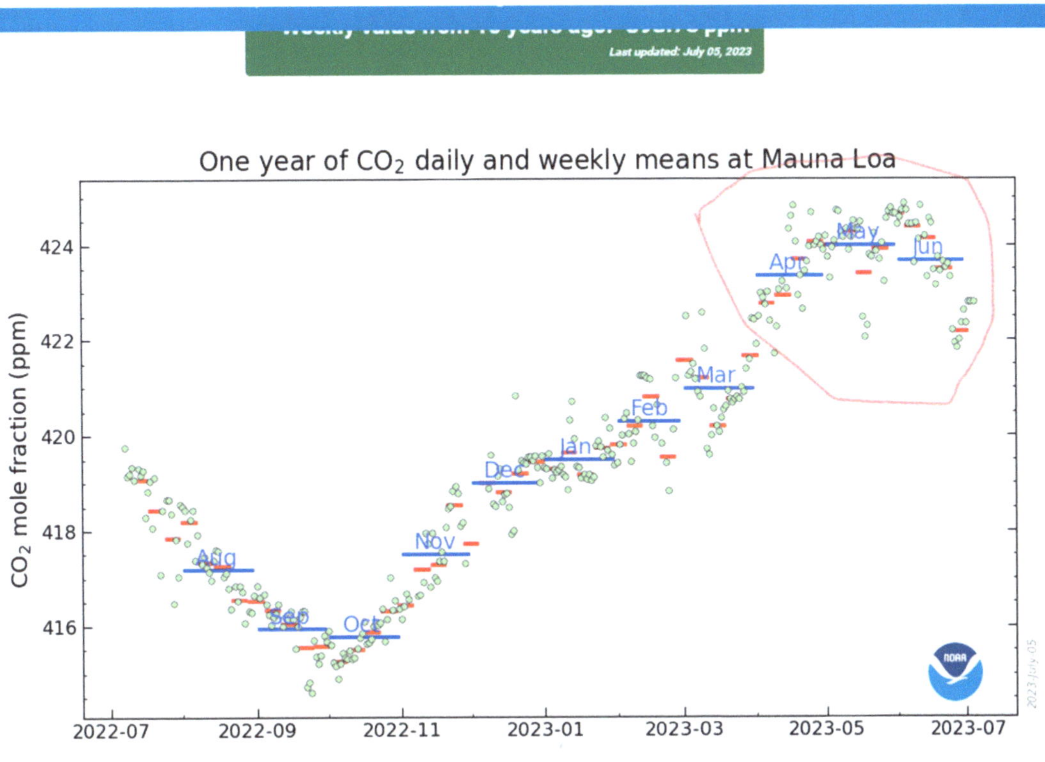

Preliminary weekly (red line), monthly (blue line) and daily (green points) averages at Mauna Loa for the last year.

Graph 6.5 2023 Peak similar to last year, in March 2023 with 35 billion trees planted.

In September the NOAA carbon dioxide data decreases. However, NOAA Mauna Loa carbon dioxide scientists didn't like how twenty-eight billion trees were lowering carbon dioxide levels. They committed more fraud (Graph 6-4) by holding the value above 416 for a week or so. Then in 1 day it dropped 1.58 ppm and has stayed low.

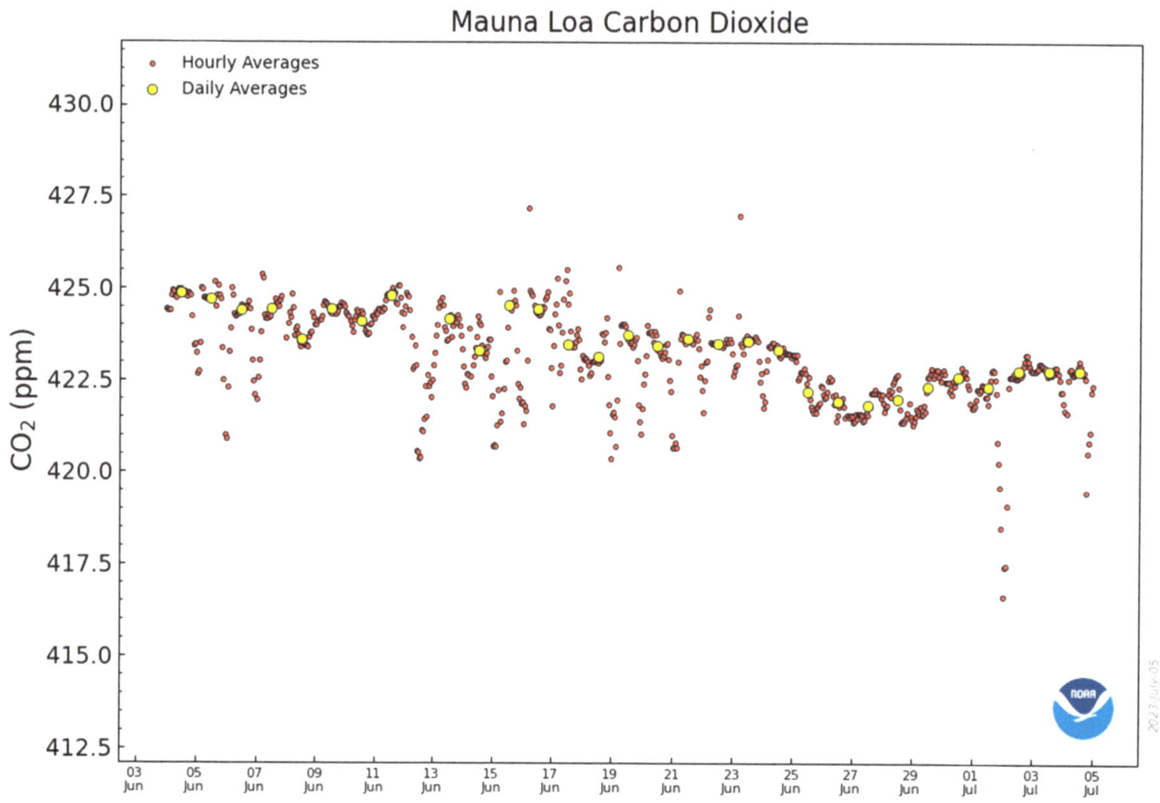

Hourly (red circles) and Daily (yellow circles) averaged CO_2 values from Mauna Loa, Hawaii for the last 31 days.

Graph 6.6

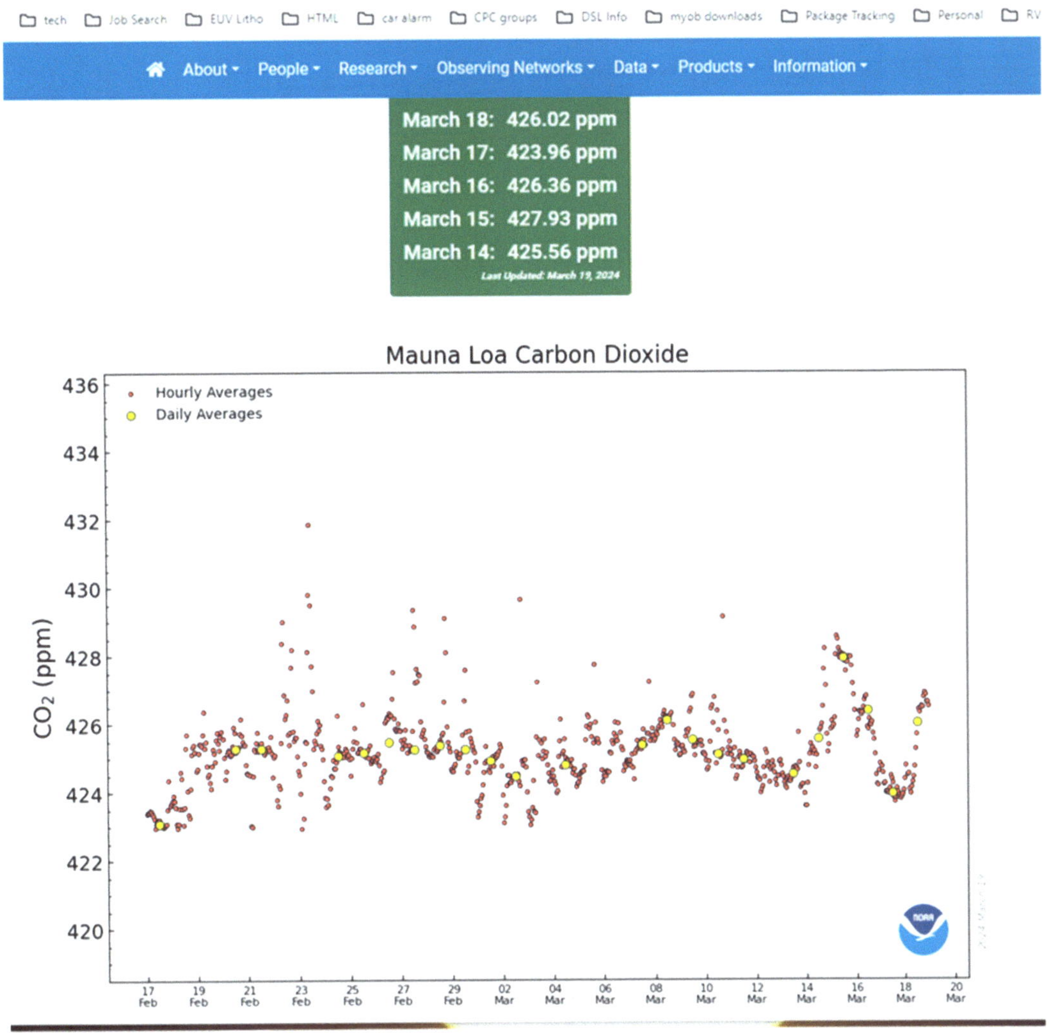

Graph 6.7

More fraud March 2024, Red arrows are more manual adjustments up. More than 43 billion native trees planted on cctruth.org are lowing atmospheric carbon dioxide quickly.

Graph 6.6 is Monthly data from June-July 2023. For 70 years the peak in carbon dioxide data was late May to Early June. This year because of Fraud the fake peak was June 13th. Then it started going down without manual adjustments. Then on June 25th NOAA started manual adjustments again causing flat carbon dioxide concentration instead of decreasing as it normally did for 70 years.

Why the NOAA carbon dioxide data are fake:

NOAA uses the flask method. They pump carbon dioxide air into the flask with water in it. Then see how much the concentration is in the water.

There are two major problems with this method:

1. The pressure from the pump varies from 1.2 times ambient pressure to 1.5 times ambient pressure.

2. No temperature control in the room and the temperature may vary +/- 2 F

The diffusion they are using is –D(dc/dx). dc is the change in concentration, and dx is the change in distance.

D is the diffusion coefficient.

For a gas, D is affected by temperature and pressure.

D(CO2) = D0[T/Ts-1]m where D0 = 13.942 × 10–9 m2/s, Ts = 227.0 K, and m = 1.7094.

Therefore, the D is a first order affected by Temperature as T/Ts-1. Ts is constant at 227 Kelvin.

https://www.sciencedirect.com/science/article/abs/pii/S00167037130022

Pressure effect is small but not negligible.

Dave White talked with NOAA Dr. Kathryn McCain of NOAA's group about the flask method. She pointed Dave to a PDF of their specification. Dave found several technical issues with their measurements, especially with control of temperature and pressure. Dr. Kathryn McCain was fired a short time later.

He also talked with Dr. Jennifer Carney, Group Leader of NIST Carbon Dioxide Measurements and Reference Materials. She said they have a team working on a standard to match WMO X2019 standard reference to NIST standard reference. She is going to talk to the team and see if Dave can join.

Chapter 6 Questions:

1. Why is carbon dioxide lowering faster than previous time?

2. How is NOAA Mauna Loa committing fraud?

Chapter 7

NiCE Fix for Southeast USA Storms

For thousands of years a warm, shallow, opaque, 26,000 km² freshwater lake would appear in the northeastern Sahara Desert (Figure 7-1) from July to October. This lake was created by the torrent of water flowing down the Nile river into modern day Egypt, submerging the flood plains and delta under up to eight meters of water.

This lake created clouds which cooled the western Africa shore water. Storms in the SE USA originated from this water (see figure 7-2). However, the Aswat Dam was installed on the Nile River in the 1960's, which stopped the flooding as in Figure 7-1.

In 2019 scientists found this solution and installed another dam downstream of this valley. The valley is now flooding as it was prior to circa 1965. Therefore, as of late September 2022, the southeast storms in the United States have stopped coming in. They are proceeding up the Atlantic Ocean, currently close to the United States but not coming ashore.

Figure 7-1 Valley to flood from the Nile River.

Dave White was on a zoom meeting with the lead scientist along with NASA and other interested scientists. They showed the dam, which was being used to backfill the valley.

The warmer the water off the coast of Africa, the further west the hurricanes travel. Prior to 1970 (when the Aswan Dam was built), the water was cooler. Most storms went straight up the Atlantic Ocean. However, after the Aswan Dam was built, the floods and resultant clouds did not appear. This caused the ocean water off the west coast of Africa to be warmer. When the water is warmer, the storms go westward to the United States. When the water is colder, the storms travel up the Atlantic Ocean until they fizzle out in the colder water of the North Atlantic.

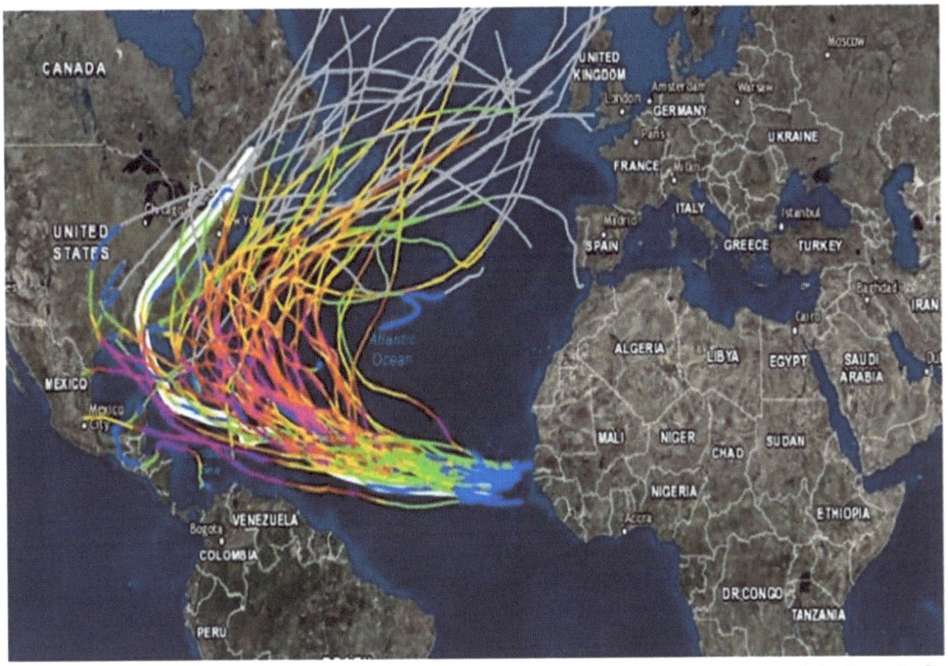

Figure 7-2 Storm track and origin of Southeast United States storms.

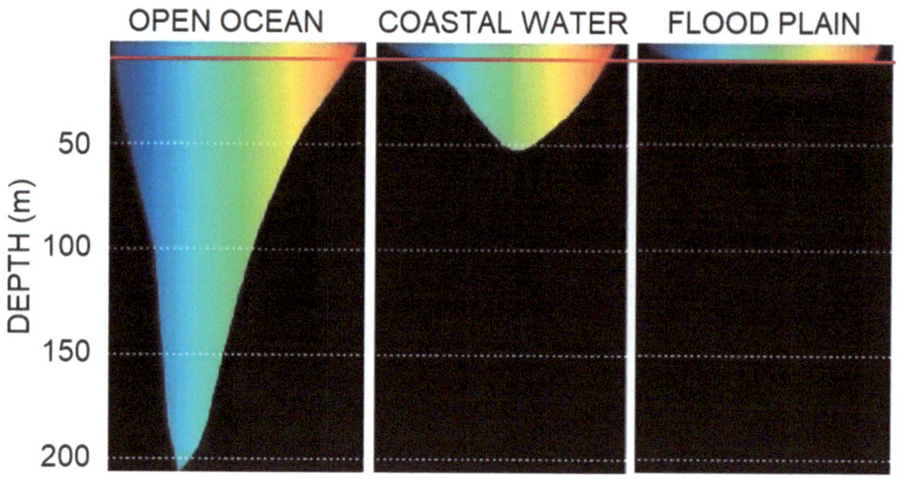

Figure 7-3 Shallower water heats up faster (bulk) and therefore evaporates faster.

More surface area evaporation occurs from shallow water than deep water,

Less and less rain falls from clouds every year. Figure 7-4

This is because there is no flooded area.

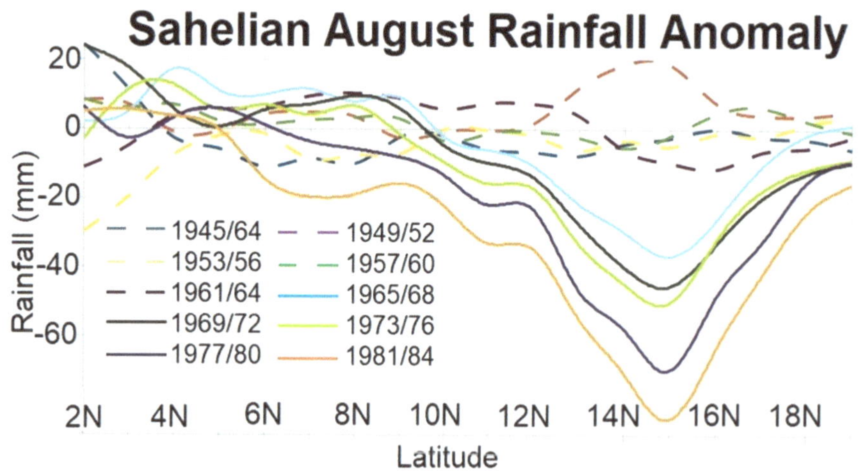

Figure 7-4 Sahelian Rainfall anomaly because of Aswat Dam installation.

The NiCE Chain Reaction

Nile Flood

. . . .

Evaporation onto Ethiopian Highlands

. . . .

Kiremt rainfall event

. . . .

African easterly waves

. . . .

Equatorial Atlantic marine stratocumulus cloud

. . . .

Summer sea isolation

. . . .

Ocean heat transport

. . . .

Additional heat released into atmosphere

. . . .

Southeast storms will stop by 2027! The weather channel reports all southeast United States storms are going up the Atlantic Ocean. The NiCE effect is occurring. Three years ago, a zoom meeting showed how the dam would backfill the delta. They said it would take effect by 2026 or 2027.

Hurricane Fiona was the "first" of the 2022 season. This was very late to be a first. Normally, the first would have been in August or early September. Remember Hurricane Katrina? The track below was affected by fifty years of NOAA models which show the hurricane must make landfall.

Puerto Rico experienced only heavy rains and flooding, as well as a temporary loss of power. The news media made this out to be another Katrina; however, it was not. Fiona did not hit the NE USA or SE Canada. The new model had it going out to sea by the time it passed by Maine.

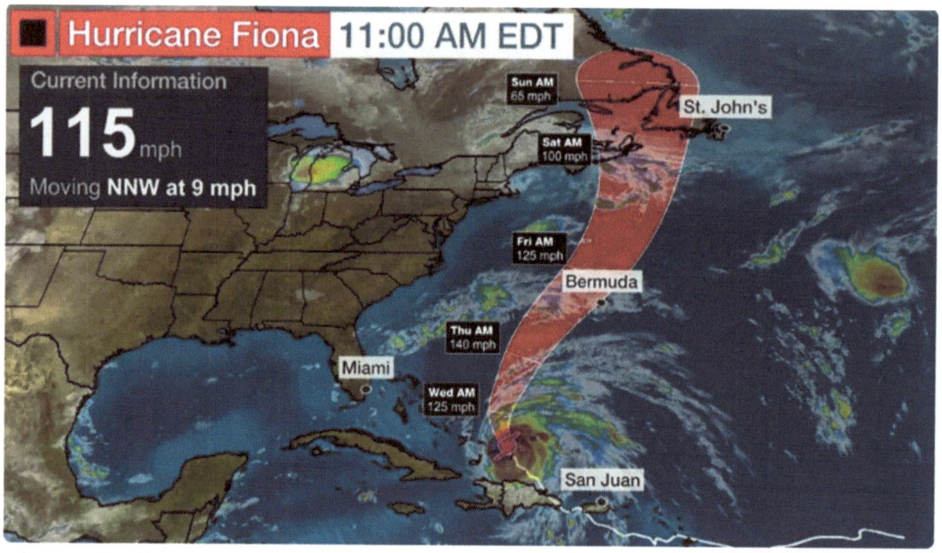

Figure 7-5 Fiona was on the wrong track according to previous NOAA models.

The NOAA models are simply wrong. They are constructed from fifty years of storm tracking such as seen in figure 7-5. Their models, based on past data, are inaccurate. A new paradigm has developed. NOAA must change its models for the new paradigm.

See Figure 7-6 below. The X is the track predicted by the outdated model. Hurricane Gaston is in the mid-Atlantic (predicted by the new paradigm), but it fizzled out as it approached colder North Sea water.

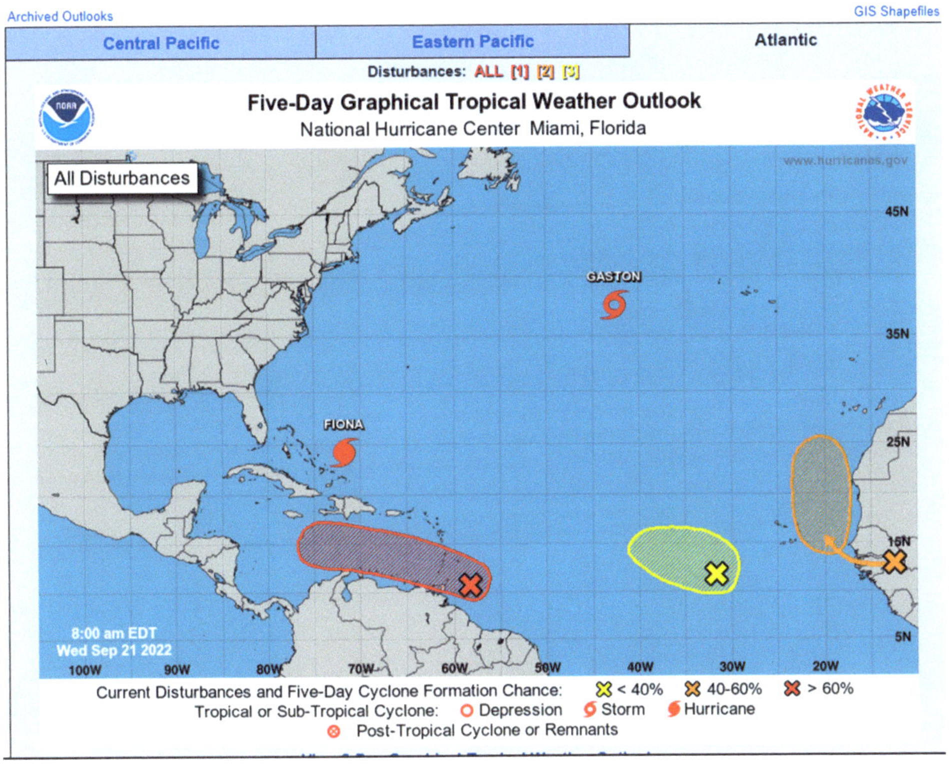

Figure 7-6 Late September 2022 storm track.

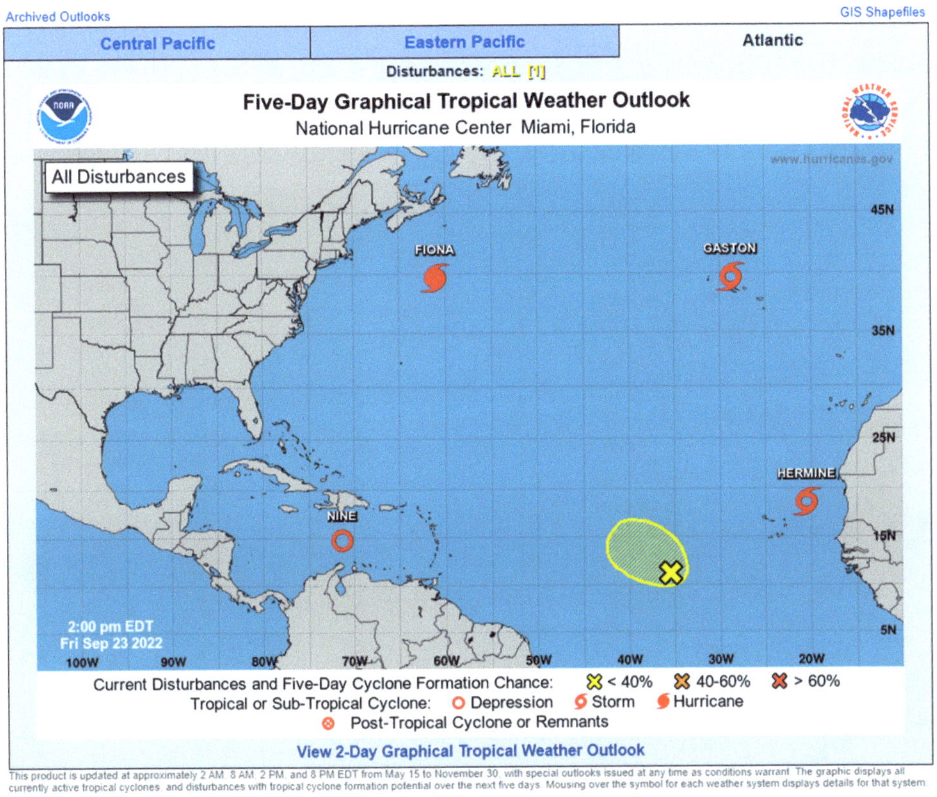

Figure 7-7 Fiona doing exactly what the new paradigm says.

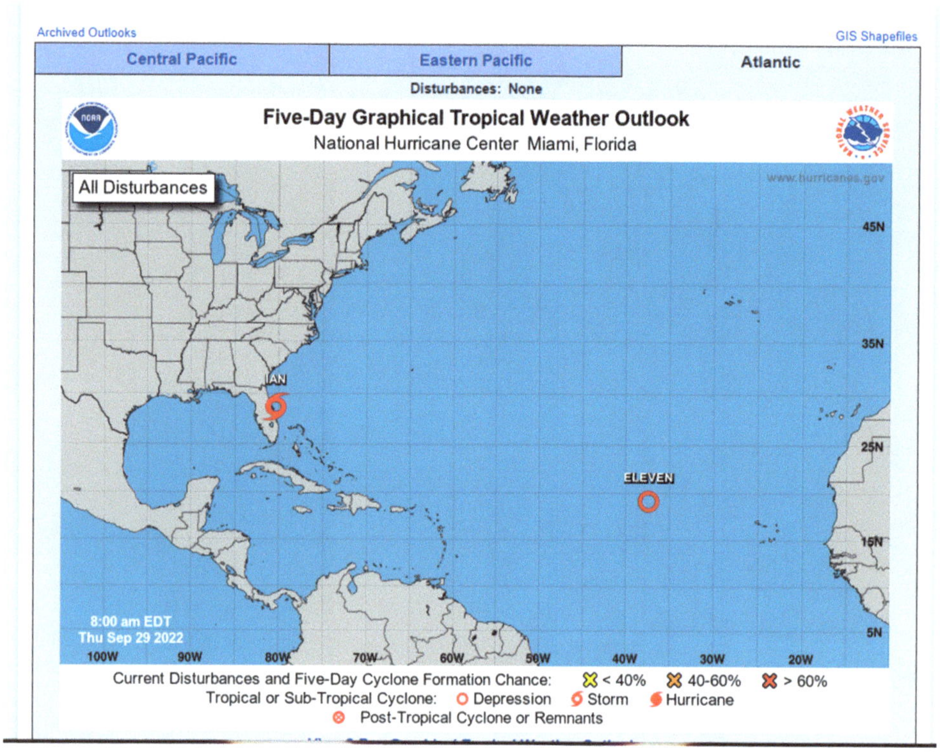

Figure 7-8

Tropical storm IAN came from South America, not from the west coast of Africa. It came north and hit the tip of Cuba, then made a right turn straight across Florida. We predicted it would keep moving northeast because of the atmospheric river created by the NiCE effect. Depression 11 is moving northward by the same atmospheric river. Our prediction was correct.

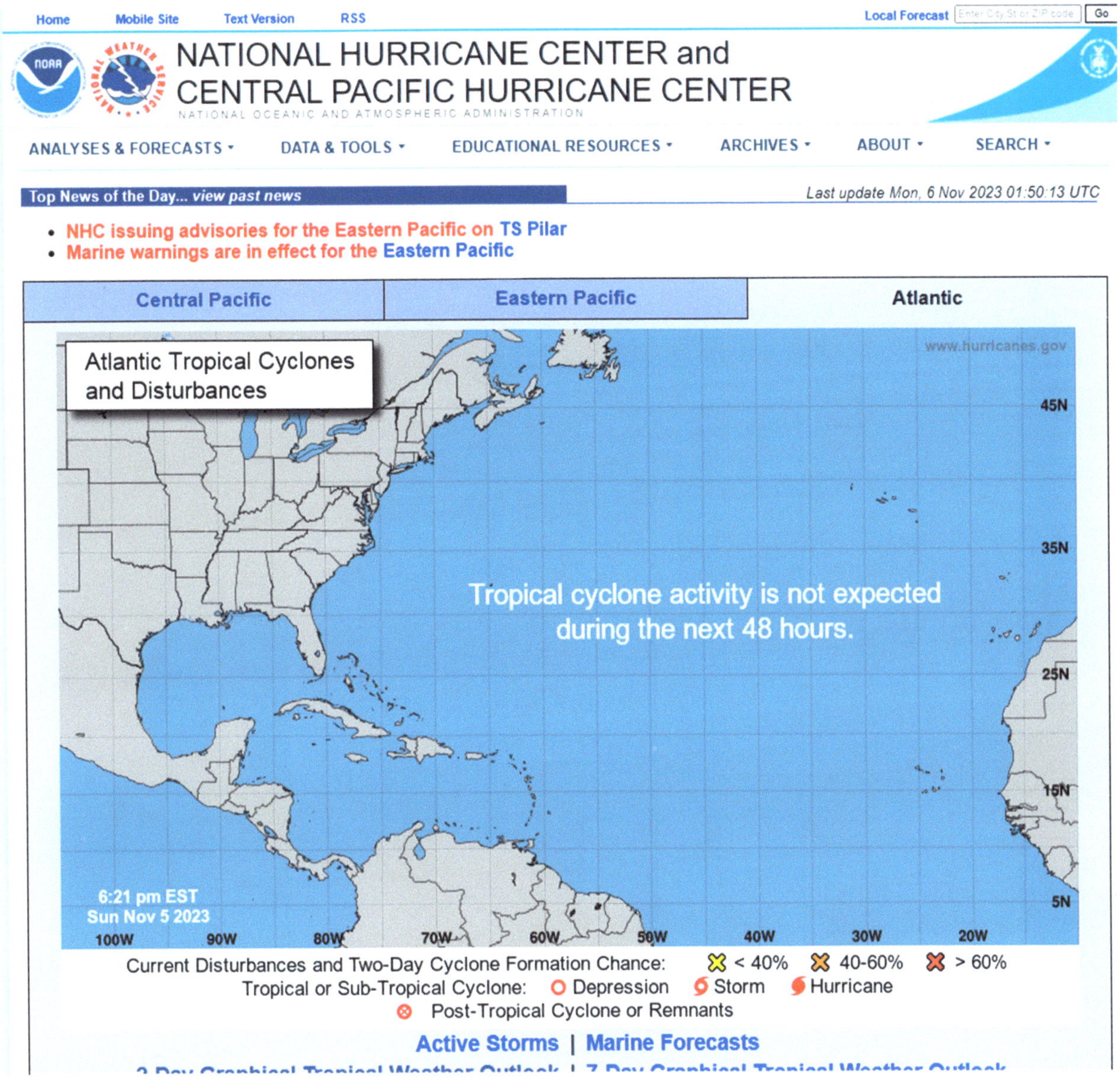

Figure 7.9

As of end of October 2023 the image from the National Hurricane Center has text which shows no expectation.

Chapter 7 Questions:

1. Why did the Aswat Dam on the Nile River have such a profound effect on storms coming into the Southeast United States?

2. How is the previous NOAA model wrong and outdated?
3. What could be a proposed solution for storms coming from South America?

References:

https://journals.ametsoc.org/view/journals/clim/26/20/jcli-d-13-00037.1.xml

https://www.nhc.noaa.gov/refresh/graphics_at2+shtml/113603.shtml?tswind120

https://www.nhc.noaa.gov/

Chapter 8

Fictional Global Sea Rise

The future of New York City? NO!

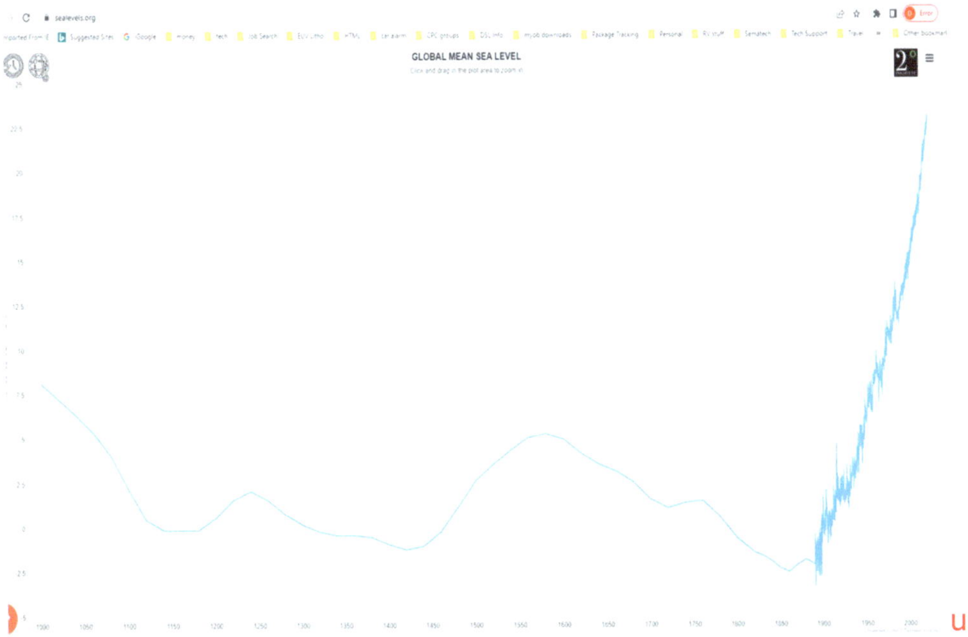

Figure 8-1 Two different measurement systems: tide gauges are light blue; satellite is dark blue.

The NOAA sea level measurements are not reliable. The NOAA satellite, Jason 3, has a minimum resolution of 25 mm. This is like using a yardstick with no gradations and putting an unfamiliar object twenty feet away, then performing measurements. The data would not be reliable. The Jacobson Glacier in Greenland has grown for the fifth year in a row. The newest satellite, the Jason 5, has a minimum resolution of 1 mm. However, they calibrated it with the unreliable Jason 3 data. Therefore, the Jason 5 data is unreliable. They should have calibrated it with the tide gauges. These are accurate, reliable, and all over the world.

❖ <u>3 Sources - Different Metrics</u>

Source	Rate	Trend	
Tide Gauges	1.4mm/yr.	<u>Linear</u>	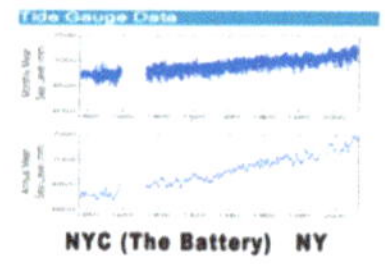
Topex/Poseidon/ Jason 1 - 3	3.3mm/yr.	<u>Linear</u>	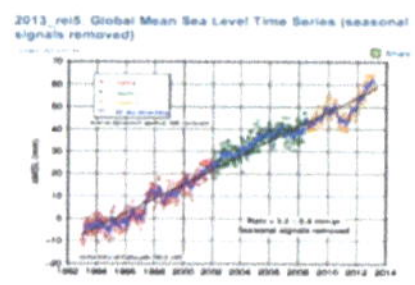
ENVISAT	0.5 - 2.5mm/yr.	<u>Linear</u>	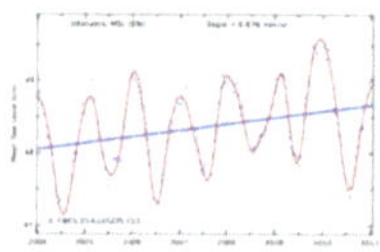

Figure 8-2 Three sources

Tide Gauge Distribution

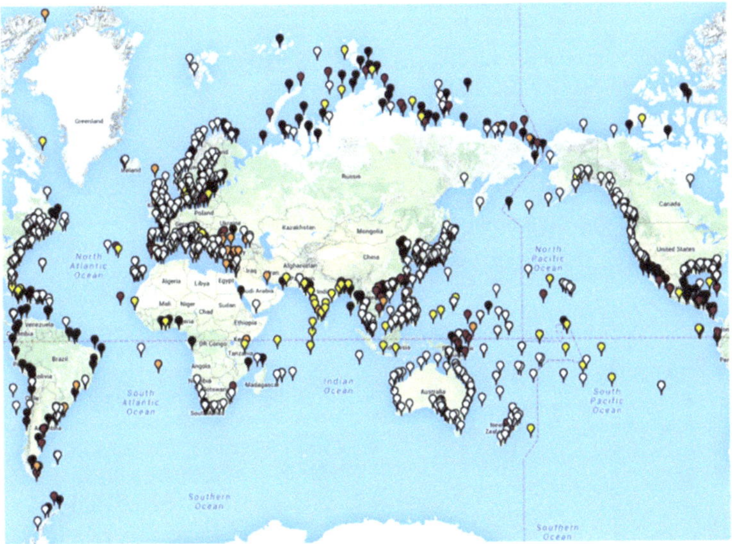

Figure 8-3 Tide gauge distribution throughout the world.

Massive 38% CO_2 Increase Since 1880

Any sign of Sea-Level Rise Acceleration? They're all linear.

In 2021 ninety of NOAA's tide gauges were chosen at random. All tide gauges look like this:

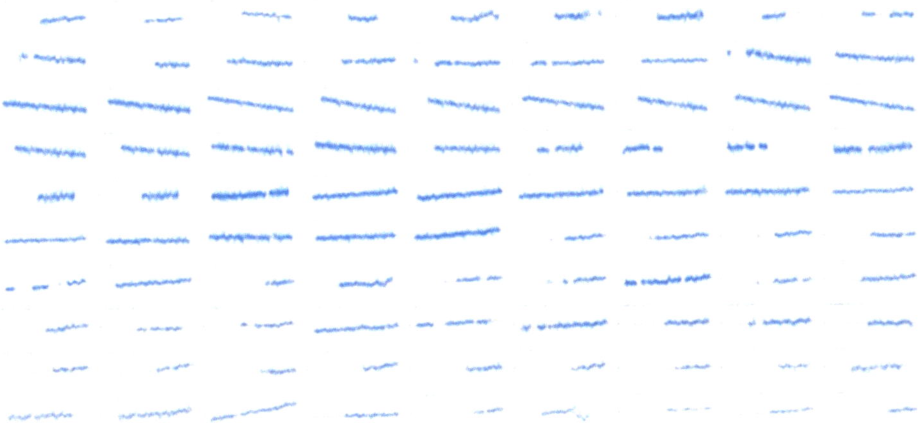

Figure 8-4 NOAA tide gauge graphs.

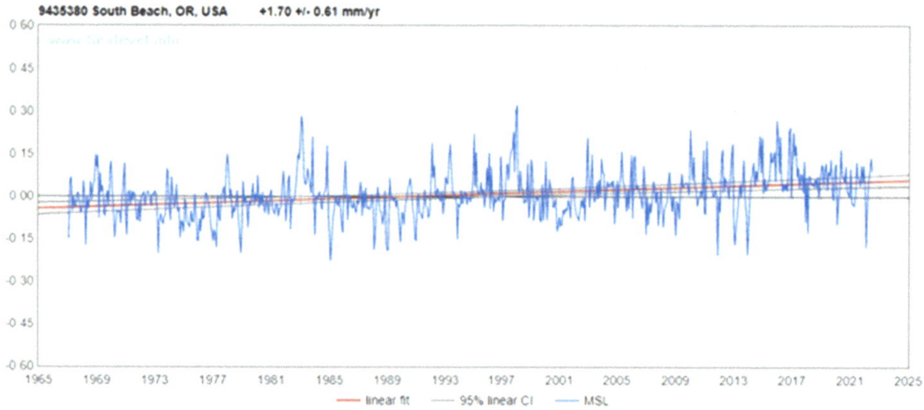

Figure 8-5 Tide gauge 15 miles south of Depot Bay Oregon.

The NOAA graphs are below. They show the same thing.

SOUTH BEACH

Station Information

Station ID	1196
Latitude	44.625
Longitude	-124.041667
GLOSS ID	157
Coastline code	823
Station code	16
Country	UNITED STATES
Time span of data	1967 – 2021
Completeness (%)	100
Date of last update	10 Feb 2022

Green Arrow	Current Station
Yellow Marker	Neighbouring RLR Station
Red Marker	Neighbouring Metric Station

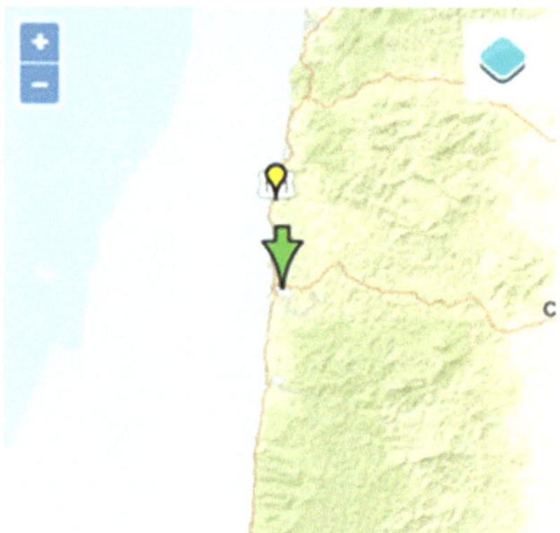

Sources: Esri, Garmin, HERE, NGA, NPS, USGS

Please note: In many cases, the station position in our database is accurate to only one minute. Thus, the tide gauge may not appear to be on the coast.

Tide Gauge Data

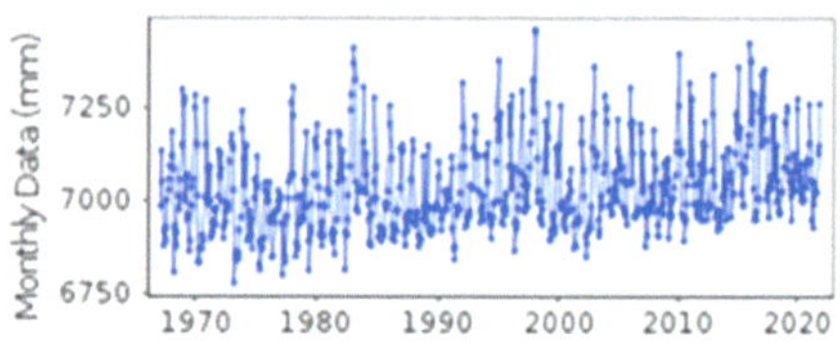

Link to larger image of monthly data plot
Download monthly mean sea level data

Figure 8-6 NOAA data online agrees with manual collected data.

There's a short answer to that question, and a much, much longer explanation. The short answer is that the very best sea-level measurement record in the world is, in my opinion, the tide-gauge measurement record from Honolulu:

https://sealevel.info/MSL_graph.php?id=Honolulu (interactive version)

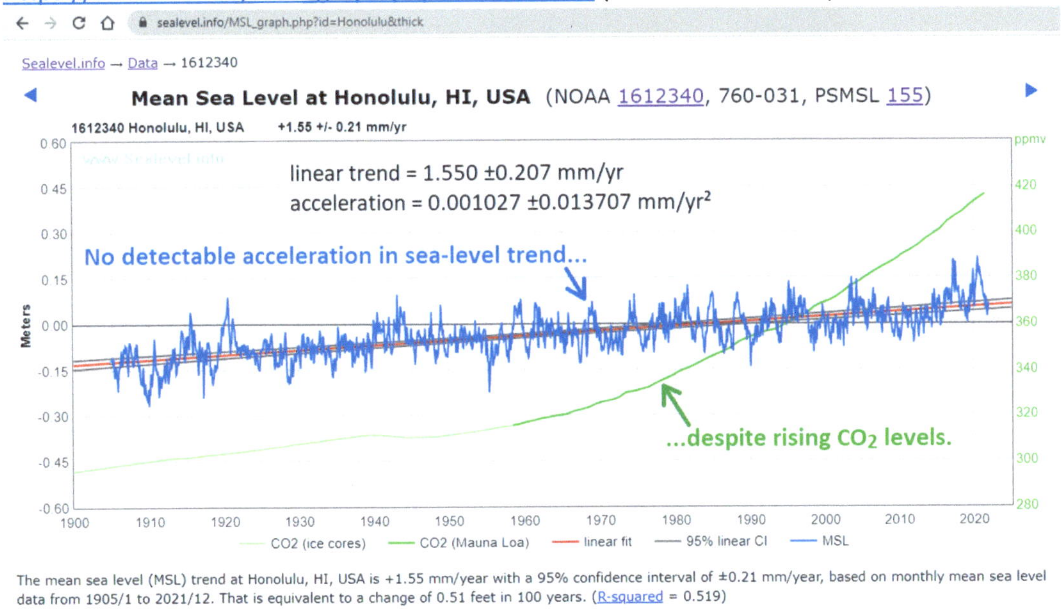

(click image to enlarge it, or see NOAA's version, here)

Figure 8-7 Honolulu Hi. Sea level raise data.

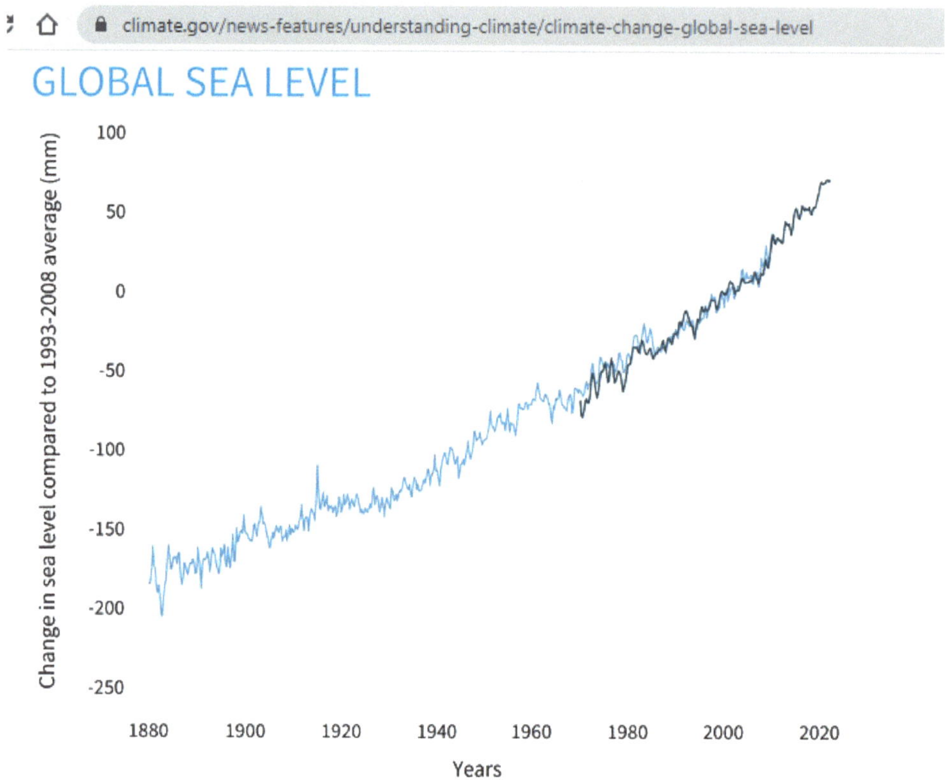

Seasonal (3-month) sea level estimates from Church and White (2011) (light blue line) and University of Hawaii Fast Delivery sea level data (dark blue). The values are shown as change in sea level in millimeters compared to the 1993-2008 average. NOAA Climate.gov image based on analysis and data from Philip Thompson, University of Hawaii Sea Level Center.

(click image to enlarge it)

Figure 8-8 Global Sea Rise data from tide gauges (light blue) and Satellite (dark blue)

• Honolulu has 117 years of continuous measurements, from the same location, with no gaps.

• Honolulu's location near the middle of the Pacific is nearly ideal.

The Pacific Ocean "sloshes east" during El Niños, and "sloshes west" during La Niñas, but Hawaii is near the "teeter-totter pivot point," so, unlike the other long Pacific measurement records, Honolulu is scarcely affected by it. In contrast, look how strongly sea-level is correlated with ENSO at Kwajalein (in the western Pacific) and San Diego (in the eastern Pacific):

https://sealevel.info/1820000_Kwajalein_San_Diego_2016-04_vs_ENSO_annot4.png

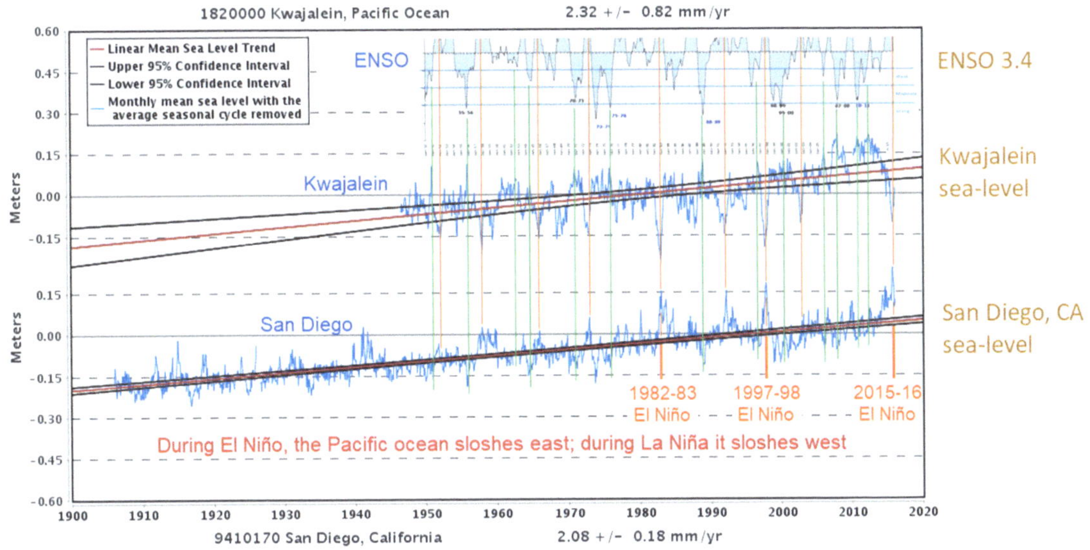

(click image to enlarge it)

Figure 8-9 Pacific Ocean sloshing.

• Oahu experiences only small tides, and almost no vertical land motion (though it does move horizontally, about 3" to the northwest, each year). Oahu is an old, tectonically stable island (several million years old, which is about 4× the age of the Big Island); the volcanoes on Oahu are believed to have been inactive for well over a million years. So Honolulu's sea-level trend is typical (about the same as the global average, i.e., about +1½ mm/year = 6 inches/century), and it is not appreciably distorted by vertical land motion.

• Unlike the north Atlantic and Baltic, Hawaii is not appreciably affected by Greenland's slowly changing gravity field.

There are other sites with measurements going back even further than Honolulu's, but all of them have either gaps in the record, or less-than-ideal locations. In particular, the Dutch have done a superb job of measuring sea-level for a very long time (for obvious reasons), but The Netherlands is close enough to Greenland that if Greenland's ice melt were to accelerate (which is the potential source which is most plausible to cause sea-level rise acceleration) the Dutch would see less than average sea-level trend change. So, even though their measurement records are very, very good, they aren't the best records to use for detecting possible acceleration from Greenland meltwater. Still, here's a photo of a Dutch dike, next to a farmhouse, and a Dutch sea-level measurement record:

https://sealevel.info/Dutch_dike_vs_Harlingen_sea-level_trend_1880x940_v06.png

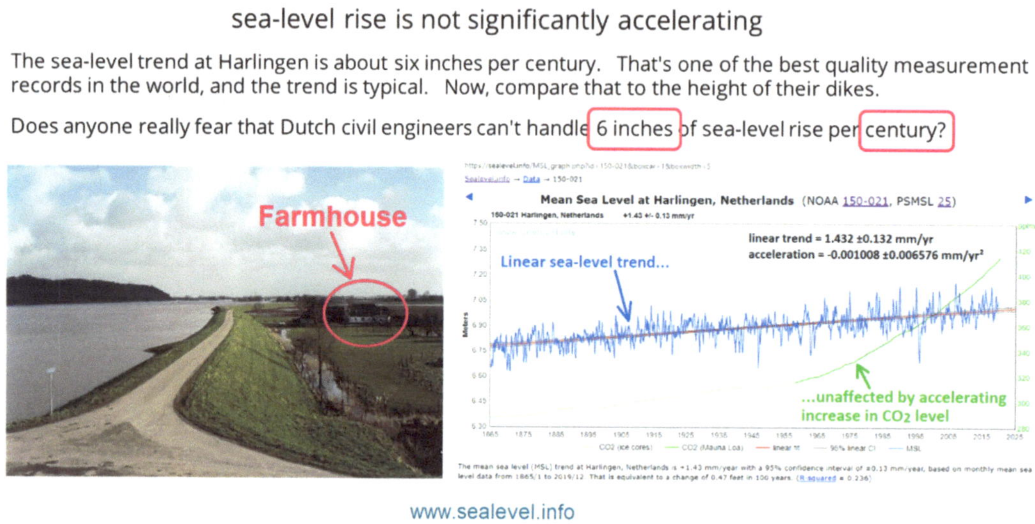

(click image to enlarge it)

Figure 8-10 The Dutch don't care about Sea level rise why should we?

- Unlike the U.S. east coast, Hawaii is not affected by variations in the AMOC.

Over the last decade, the southeastern U.S. coast, from Florida to North Carolina, has seen noticeable regional sea-level acceleration, where the Gulf Stream (part of the AMOC) skirts the coast. (It hasn't appreciably affected New York.) The cause is apparently a (presumably transient) acceleration of the AMOC.
https://sealevel.info/Gulf_Stream_wikipedia_200pct_with_NYC_Wilmington_Key_West_labeled.png

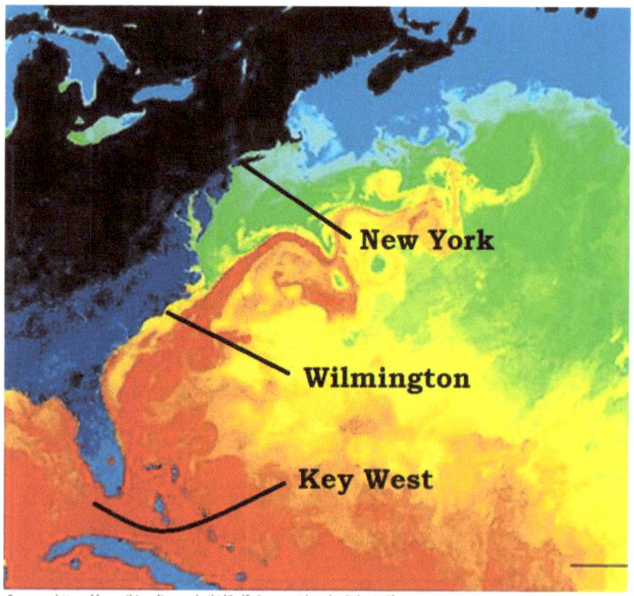

Source: https://en.wikipedia.org/wiki/Gulf_Stream#/media/File:Golfstrom.jpg

(click image to enlarge it)

- The highly linear trend measured by Honolulu's tide gauge is typical of most coastal measurement records, which have seen little or no acceleration for at least nine decades. The amount of measured acceleration reported in long tide-gauge measurement records has varied from negligibly positive to negligibly negative. This table from Houston (2021) summarizes the findings of ten of the best such studies:

https://sealevel.info/Houston2021_Table2_annot1.png

Table 2. *Comparisons of results in earlier studies with those of this study.*

Study	Number of Gages	Based Completely on Measurements	Record Lengths Years	Period Years	a_2 mm/y^2
Douglas (1992)	23	Yes	80	1905–85	−0.011 ± 0.012
Holgate (2007)	9	Yes	100	1904–2003	−0.0012
Jevrejeva *et al.* (2008)	1023	No	Average 30	1700–2000	0.01
Wenzel and Schröter (2010)	56	No	56	1950–2006	0.0016 ± 0.0043
Houston and Dean (2011)	23	Yes	105	1905–2010	−0.012 ± 0.012
Church and White (2011)	290	No	Average 55	1880–2009	0.009 ± 0.003
Ray and Douglas (2011)	89	No	>40	1807–2009	0.00
Olivieri and Spada (2013)	315	Yes	>50	1810–2010	0.0042 ± 0.0024
Hogarth (2014)	117	No	>100	1807–2012	0.0105 ± 0.0081
This study	149	Yes	>75	1807–2019	0.0128 ± 0.0064

Houston, J.R., 2021. Sea-level acceleration: Analysis of the world's high-quality tide gauges. *Journal of Coastal Research*, 37(2), 272–279. doi:10.2112/JCOASTRES-D-20-00101.1

Figure 8-11 Sea Level acceleration is very small.

Note the pink box around the greatest acceleration reported by any of those studies: 0.0128 ±0.0064 mm/year^2. A sea-level acceleration of 0.0128 mm/year^2 continued for 100 years would add a negligible 64 mm **(2.5 inches)** to sea-level — obviously not worrisome.

Take a closer look at that climate.gov graph, and you'll notice that it is actually two different graphs, superimposed. One is from a 2011 paper by Church & White, based in tide gauge analysis. The other, which I'll call the "Thompson trace," is described as "University of Hawaii Fast Delivery sea level data," but it apparently is not from the "Fast Delivery" tide gauge data at the linked web page. Rather, it is from unpublished work "based on analysis and data from Philip Thompson, University of Hawaii Sea Level Center."

The actual data sources are not revealed anywhere on that climate dot gov page, but if you go to the University of Hawaii Sea Level Center web site, you can find this:

Data sources: Trends were calculated using (1) tide gauge data from the UHSLC Fast-Delivery database, and (2) Ssalto/Duacs altimeter products that were produced and distributed by the E.U. Copernicus Marine and Environment Monitoring Service (CMEMS). The latter are identical to the products formerly distributed by Aviso.

In other words, the Thompson trace conflates sea-level measurements from tide gauges (from what appears to be a 1969 starting date to 1993), with satellite altimetry data (which began in 1993). It is not revealed whether the portion of the Thompson graph from 1993 to present is exclusively satellite altimetry, or some unspecified blend of satellite altimetry and tide-gauge data, but, either way, the graph is a misleading "apples vs. oranges" mashup.

Note that satellite altimetry can only measure sea-level of the open ocean, it cannot measure it near the coasts. Also, in most respects, the satellite altimetry measurements are of much lower quality than tide gauge measurements.

Conflating coastal tide gauge and satellite altimetry measurements is a big mistake, because they are two very different types of measurements, made by completely different means, in entirely different places, of "sea-levels" that are affected by different things, and there's no reason to expect them to be equivalent.

Splicing those two different kinds of data together into one graph, with one kind of data used for the left part of the graph, and the other kind of date used for the right part of the graph, with no indication that you've done so, is scientific malpractice. It's what one wag called "the IPCC sea level Nature trick."

Hansen & Sato also splice the two kinds of data, but at least they use contrasting colors:
https://sealevel.info/Hansen_Sato_SL.1900-to-2021-08_downloaded_2022-02-04.png

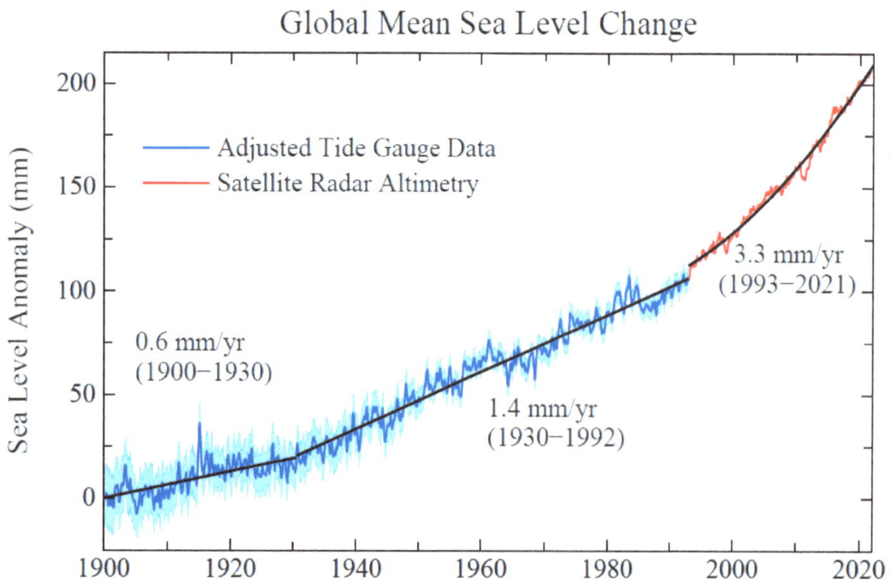

Source: http://www.columbia.edu/~mhs119/SeaLevel/SL.1900-2021.pdf (Makiko Sato & James Hansen) 2021/08/31

Figure 8-12 showing sea level rise graph with two systems. Both linear. However, most graphs online use a single color and remove the legend.

Note, too, that climate.gov started the Thompson trace (in 1969) at an offset from the Church & White trace. It shows a larger trend than the Church & White trace, and they picked a negative offset which caused the other end of the Church & White trace to be covered by the Thompson trace, making the fact that 12 years of data are missing from the Church & White trace less obvious.

We added the orange annotations:

https://sealevel.info/climate_dot_gov_misleading_global_sealevel_200pct_graph4.png

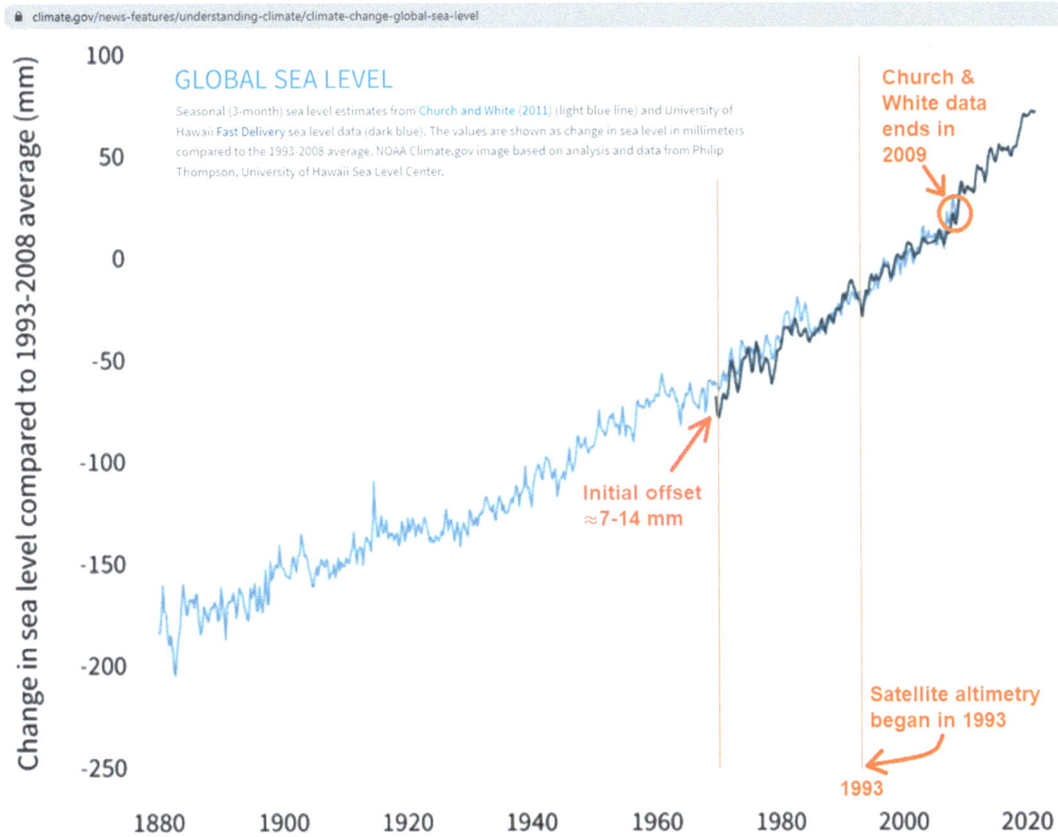

Figure 8-12 More information on the unnecessary switch from Tide Gauges to Satellite altimetry

It gets worse. Not only is the climate.gov graph actually a composite of two different incompatible graphs, and not only is the Thompson trace a spliced composite of two entirely different kinds of data (satellite and tide gauge), the satellite altimetry data, itself, is also a composite of many short and strikingly different measurement records.

Here's what the individual measurement records looked like back in 2012. Notice how different they are from one another.

https://sealevel.info/MSL_Serie_ALL_Global_IB_RWT_NoGIA_Adjust_2012_cropped.png

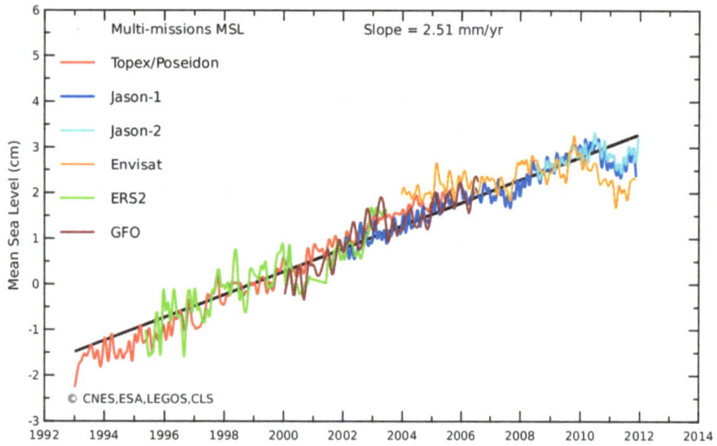

Figure 8-13 Satellite Altimeter data.

But then they "corrected" a decade's worth of Envisat data (the orange trace above):

This graph shows the before and after versions of the Envisat data:

https://sealevel.info/envisat_msl_correction_from_esa02.png

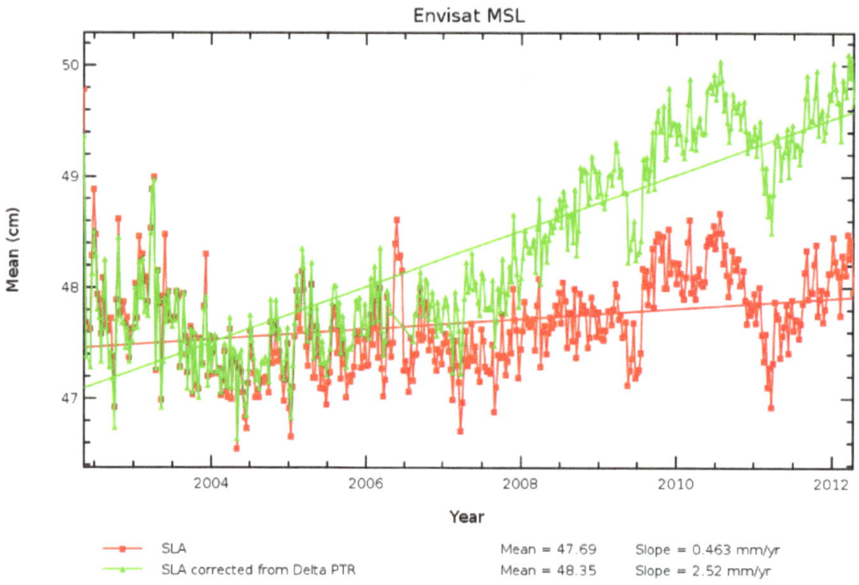

Source: https://earth.esa.int/web/guest/content/-/article/improvement-of-envisat-ra-2-reprocessed-data-v2-1

Figure 8-14 Envisat Mean Sea level data.

When the various satellite records are combined into a single trace, it conveys a false sense of measurement precision. The reality is that the satellite altimetry measurements of sea-level are very rough.

Satellite altimetry measurements of sea-level trends are subject to numerous corrections and adjustments, making them unsettlingly malleable. They were showing a decelerating sea-level trend (not widely publicized, of course), but in 2014 Anny Cazenave adjusted them, making the trend suddenly linear.

This graph is from her paper. The left side is the original, showing sea-level rise slowing down; the right is "corrected" showing the trend no longer slowing.
https://sealevel.info/nclimate2159-f1_large_trimmed1.png

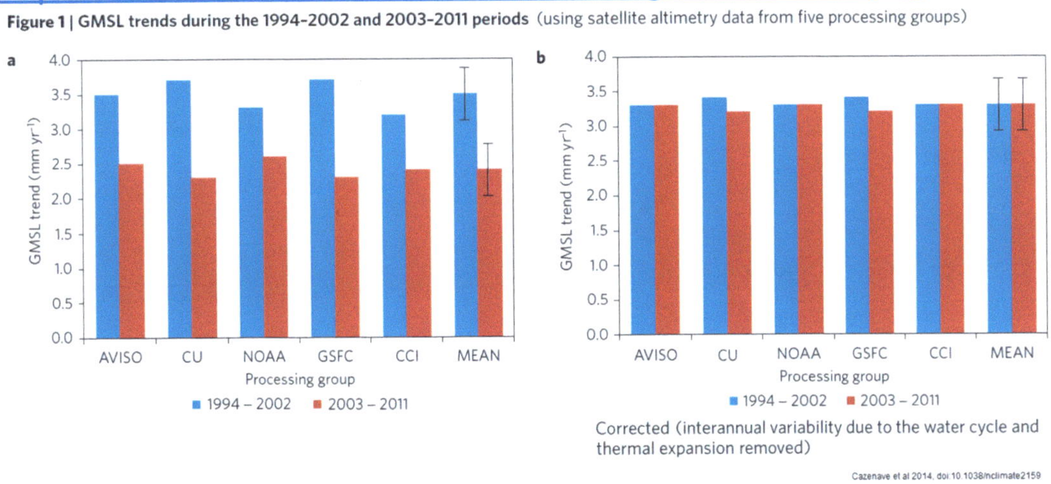

Figure 8-15 Global mean sea level trends

In 2017 the same team found additional adjustments to make, which, when combined with a few additional years of data, turned the linear trend into an acceleration. This graph is from their paper:

https://sealevel.info/2017GL073308_fig4_scaled2match2.png

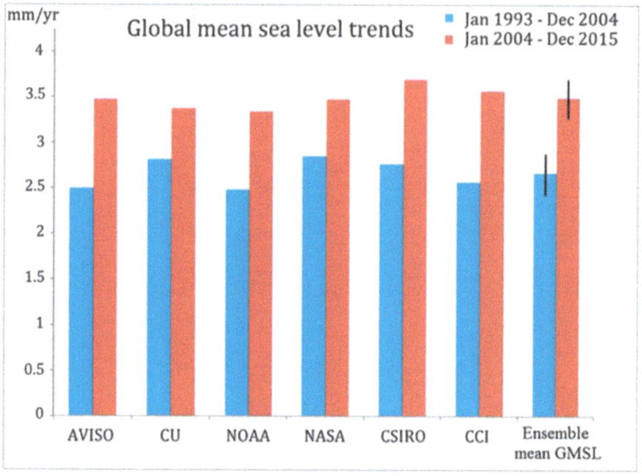

Figure 4. GMSL trends after TOPEX A drift correction computed over two time spans (January 1993 to December 2004 and January 2004 to December 2015) using satellite altimetry data from six processing groups (see details in the supporting information). The EM GMSL trend (average of the six GMSL data sets) is also shown.

Dieng, Cazenave et al 2017, doi:10.1002/2017GL073308

Figure 8-16 Global Mean Sea Level Trends

In 2018 Nerem *et al* (U. Colorado) published another paper revising satellite altimetry data. They, too, reported that satellite altimetry now shows accelerating sea-level rise, but when you read the details you discover that the "acceleration" they found was created by ***slowing*** the rate of sea-level rise "measured" by Topex-Poseidon prior to 1998. Yes, you read that right: they "fixed" 20 year-old data, thereby creating the appearance of acceleration, when the early Topex-Poseidon data is compared against the newer satellites' data.

So it is now widely claimed that satellite altimetry measurements show that the rate of sea-level rise has accelerated. We question whether any conclusions at all can be drawn from such malleable measurements, other than the conclusion that those data are untrustworthy!

You would think that the strenuous search for "acceleration" in the satellite altimetry measurements must be unique. But, believe it or not, there's a somewhat similar history behind that Church & White (2011) graph, which is the other trace on the climate.gov graph.

First, a bit of background: Most long, high-quality, individual tide gauges have recorded little or no acceleration in trend in the last ninety years or more. However, some of the longest measurement records do record a little bit of acceleration sometime before about 1930, notably including two of the longest: Brest, France (PSMSL ID #1), and [Swinoujscie, Poland] (PSMSL ID #2) The acceleration was most striking at Brest, which recorded a flat trend during the 1800s, but about 1.5 mm/year of sea-level rise since 1900; here are two graphs:

https://sealevel.info/190-091_Brest_1807-1900_vs_1900-2016.png

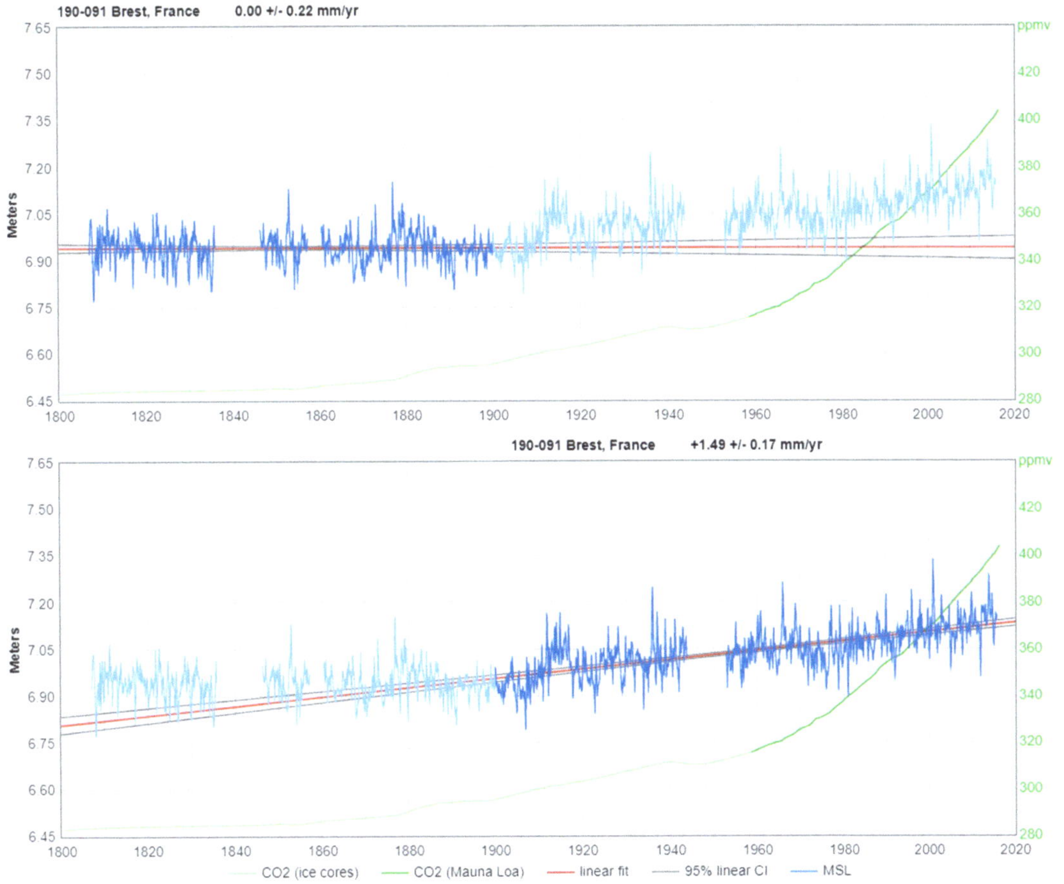

Figure 8-17

A report by Church and White encouraged the perception of sea-level rise, but a careful read of the paper found problems. Here's an interesting snippet: There are only a few tide measurement records which include much of the 1800s, so those gauges have a lot of weight in any composite index which goes back that far. So very long composite sea-level graphs tend to show a bit of acceleration, sometime between the late 1800s and about 1930. (Obviously that predates most of the anthropogenic increase in CO_2.) Keep that in mind when reading the backstory about Church & White (2011).

The story starts with the most famous sea-level paper of all: Church, J. A., and White, N. J., 2006, "*A 20th Century Acceleration in Global Sea-Level*

Rise." Geophysical Research Letters, Vol. 33, L01602, 4 PP. DOI:10.1029/2005GL024826. Church & White did quadratic regression of averaged and adjusted tide gauge data, and detected a small acceleration in rate of sea level rise for the 20th century as a whole, and found an acceleration of 0.008 ±0.008 mm/yr^2 for the 20th century as a whole, and 0.012 ±0.006 mm/yr^2 for a longer period starting in 1870.

It was the first paper claiming to have detected sea-level acceleration, and it was widely touted as confirmation that anthropogenic climate change was causing accelerated sea-level rise, but Church admitted:

"An additional spatially uniform field is included in the reconstruction to represent changes in GMSL. Omitting this field results in a much smaller rate of GMSL rise..."

In other words, their analysis didn't find enough sea-level rise, so they added a fudge factor! (BTW, I have serious reservations about their EOF-based averaging process.)

Here's another interesting sentence:

"The *rms* residual to the linear fits is lower at 5.8 mm (cf 7.5 mm), consistent with **much of the acceleration occurring in the first half of the 20th century rather than a smooth acceleration over the whole period.**"

"Much of the acceleration occurring in the first half of the 20th century" turns out to have been an understatement. I did a reanalysis of their data, and found that <u>all</u> of the acceleration was prior to 1925. Obviously that is not consistent with rising CO_2 levels being the cause.

Church and White used minimum-variance unbiased estimator quadratic regression analysis, so we stuck with their procedure. Our findings revealed that after 1925 their data showed a small deceleration in rate of sea level rise, rather than acceleration. All of the "20th century" acceleration which they had detected had occurred prior to Henry Ford selling his first Model A.

Here's a graph showing the result of starting the regression at 1925 instead of 1900:

https://sealevel.info/jnathaz1/CW06_1925-latest.png

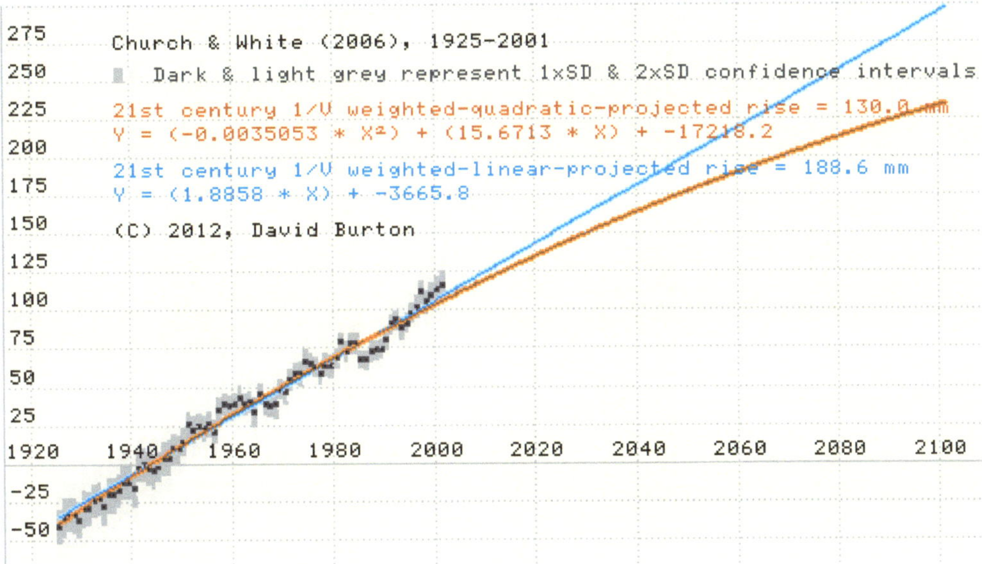

Figure 8-19

In 2009 Church & White released a new dataset on their web site, with data through 2007 (six additional years of data), apparently calculated from a different set of tide gauges. But they published no paper about it, and we wondered why not?

So we repeated their 2006 analysis, but using their 2009 data, and discovered the answer to that question. All of the 20th century acceleration was gone. Even when the quadratic regression was started at 1900 the result was a slight deceleration, and when the quadratic regression was started at 1925 the deceleration was even greater than it had been with the 2006 data:

https://sealevel.info/jnathaz1/CW09_1925-latest.png

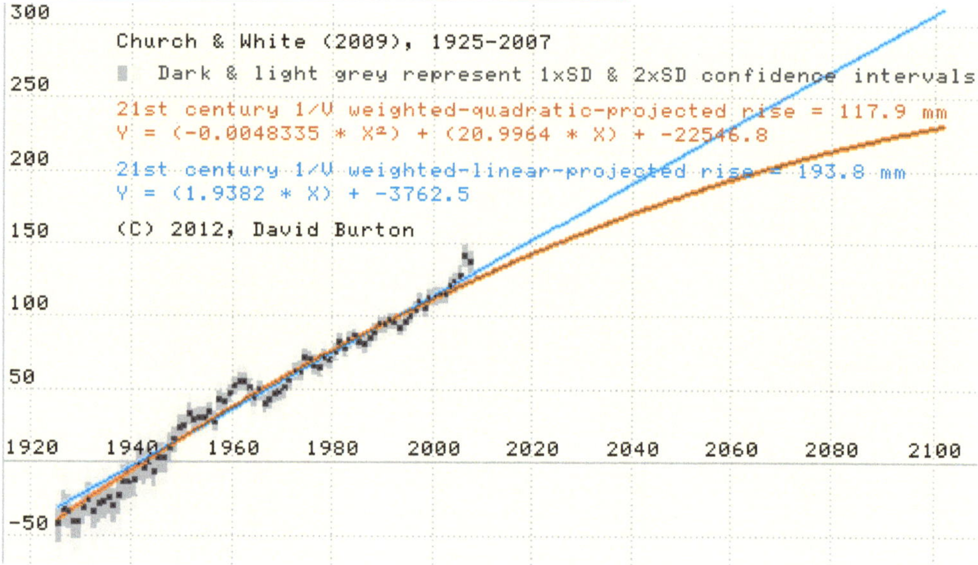

Figure 8-20

I shared my results with Drs. Church & White, and on June 18, 2010, Dr. Church replied, confirming my analysis: *"For the 1901 to 2007 period, again we agree with your result and get a non-significant and small deceleration."*

Obviously, the reason they didn't publish a paper about it is that the title would have had to have been something like, ***"Neeeever mind, no 20th Century Acceleration in Global Sea-Level Rise, After All."*** 😊

In 2011, Church and White released a third dataset, with data extended through 2009. This one showed a very slight acceleration in sea level rise after 1925, though much smaller in magnitude than the deceleration seen in their previous data sets. The post-1925 acceleration in that data set, if it continued to 2080, would add just 0.8 inches of sea level rise, compared to a linear projection.

https://sealevel.info/jnathaz1/CW11_1925-latest.png

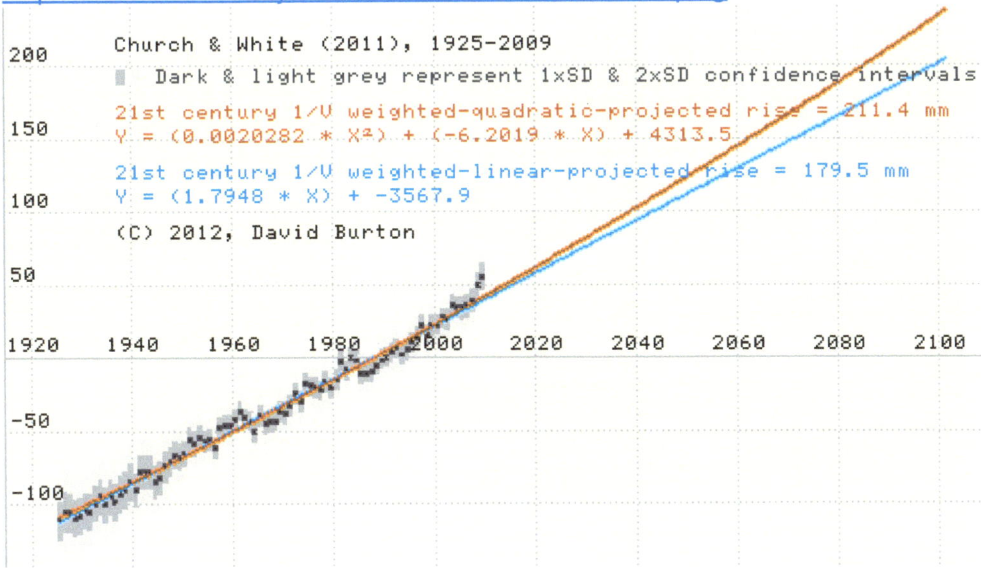

Figure 8-20

The light blue trace which climate.gov uses on their web page is from that 2011 paper.

As you can see, the selective publishing and use of only those results which support the desired narrative can significantly distort the public perception of what the scientific evidence actually shows.

In the chapter references is Dave Burton's manuscript on his findings (8-1).

So, you see, despite herculean efforts to find the widely predicted acceleration in the sea-level measurements, as a consequence to anthropogenic climate change, the best evidence is that it is negligible. Because some of the best long measurement records (Brest, Swinoujscie, and a few others) recorded a slight acceleration coincident with emergence from the Little Ice Age (probably about 1°C of warming), I do think it is plausible that we could get a similar amount of additional sea-level rise acceleration in response to an additional 1°C of warming. But the amount we've seen so far is so slight that in many of the best measurement records it is not evident at all.

In fact, the "global" (average) sea-level trend is so slight that in a great many places it is dwarfed by local factors, like erosion, sedimentation, and vertical land motion. Greta Thunberg's hometown of Stockholm is one such place:

https://sealevel.info/MSL_graph.php?id=Stockholm
https://sealevel.info/050-141_Stockholm_Sweden_1889-

2018_smoothed_vs_CO2_annot1.png

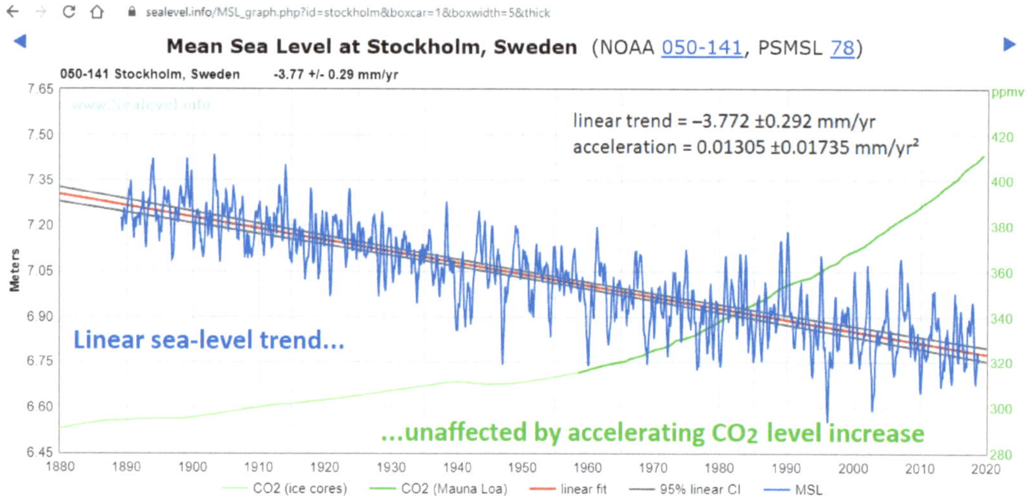

The mean sea level (MSL) trend at Stockholm, Sweden is -3.77 mm/year with a 95% confidence interval of ±0.29 mm/year, based on monthly mean sea level data from 1889/1 to 2018/12. That is equivalent to a change of -1.24 feet in 100 years. (R-squared = 0.504)

Figure 8-21

Many people have a hard time believing that global warming is not causing significant acceleration in sea-level rise. Many climate scientists assume that it simply must be so, and that assumption tends to be reflected in the models that they code. Surely warming must accelerate both ice melt and thermal expansion, right? So how can it be that sea-level rise is so insignificant, despite substantial global warming?

I think that the answer must be that warming has multiple effects, and ice melt & thermal expansion are just two of them. Of course, warming does cause melting, in some places, in the right circumstances. But the most important factor affecting annual changes in ice mass, for the Greenland and Antarctic ice sheets, is *NOT* melting. Nor is it iceberg calving, nor sublimation.

The most important factor affecting the annual ice mass balance of ice sheets and glaciers is ***snowfall.***

For both the Greenland and Antarctic ice sheets, sSnowfall is the most important factor affecting net ice mass flux, and snowfall was also certainly a very important factor affecting the advance and retreat of the great Laurentide, Fennoscandian & Cordilleran ice sheets,10K+ years ago.

A warming climate accelerates snowfall accumulation on ice sheets and glaciers, through two distinct mechanisms:

#1. Warmer air carries more moisture. Below freezing, it's 8-12% more moisture per

1°C of warming. (That's why heaviest blizzards occur when temperatures are only moderately below freezing.)

- https://web.archive.org/web/20190928155425/https://web.gccaz.edu/~lnewman/gph111/topic_units/labs_all/water%20vapor%20capacity%20of%20air.pdf
- https://www.google.com/search?q=%22too+cold+to+snow%22

Reduced sea ice coverage = more open water, increasing ocean evaporation and thus "lake/ocean-effect snowfall" (LOES) on downwind land.

- http://archive.is/vp9eP#selection-1257.0-1257.450
- https://google.com/search?q=%22lake%2Focean-effect+snowfall%22

The importance of LOES is illustrated by the story of Glacier Girl, a P-38 warbird which made a forced landing on the Greenland Ice Sheet during WWII. It was buried by snowfall which averaged ≈70 feet/yr (no, that's not a typo!), and recovered 50 years later.

That snow represents evaporated water, mostly removed from the North Atlantic and Arctic Oceans, which then fell as ocean-effect snow on the Greenland Ice Sheet.

https://p38assn.org/glaciergirl/images/GPR-measurement-concept.jpg

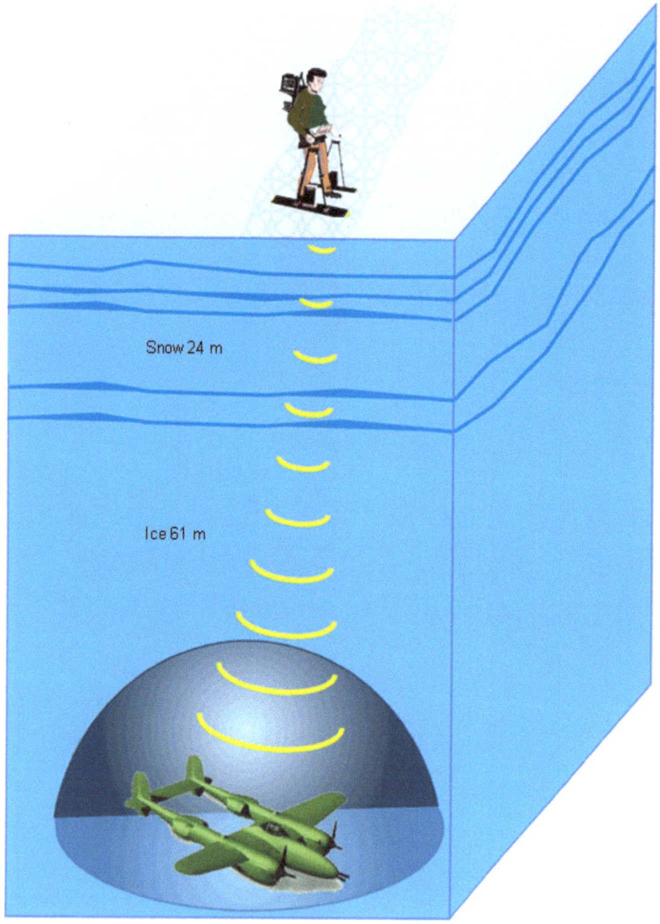

Figure 8-22 Glacier Girl

The story of Glacier Girl is fascinating. You can read more about it here:

- https://www.airspacemag.com/history-of-flight/glacier-girl-the-back-story-19218360/?all
- https://p38assn.org/glaciergirl/recovery.htm

The near-linearity of sea-level trends over the last nine decades, despite a warming climate, strongly suggest that accelerations in ice mass loss in some places are being roughly balanced by accelerations in ice mass gain in others, through increased snowfall accumulation. Here's a list of a half-dozen papers (and a few articles), published over a span of about thirty years, about accelerating snowfall accumulation in Antarctica:
https://sealevel.info/feedbacks.html#antarcticsnowcitations

Remarkably, it apparently took about thirty years (counting from this 1991 paper) for the climate modelers to discover that warmer temperatures increase snowfall accumulation on Antarctic ice sheets, largely offsetting ice loss from ice melt and iceberg calving, and to incorporate that fact into their new generation of climate models.

The bottom line is that sea-level measurement data is very encouraging. The data indicate that there's no reason to expect the large, damaging accelerations in sea-level rise which IPCC reports and many authors predict.

Chapter 8 Questions:

1. Why are the NOAA satellites' data unreliable?
2. Why are the tide gauges the correct data to use?
3. Why did NOAA use the unreliable Jason 3 satellite for calibrating the Jason 5 satellite?
4. What makes the Honolulu sea-level measurement record so good?
5. Why does graph 8-7 look so different from this heavily publicized climate.gov graph, which appears to show accelerating sea-level rise?

References:

8-1 DOI:10.1007/s11069-012-0159-8

Chapter 9

Photosynthesis Issues

The World Economic Forum has correctly stated that we need to plant 1 trillion trees, which will—in just ten years--drain the atmosphere quickly by increasing consumption of CO_2 (by 30 gt to 100 $gtyr^{-1}$).

The IPCC incorrectly asserts that the total carbon dioxide emissions are 35 billion tons of human emissions, 3 billion of which are caused by deforestation. However, the switch-over of the Amazon to an oxygen sink and carbon dioxide producer is actually 15 billion tons of additional, unaccounted for CO_2 annually.

In spite of these depressing numbers, we can have a lot of hope when we plant trees. Atmospheric carbon dioxide will lower to 330 ppm by 2031 if we can plant 200 billion trees by 2030. On cctruth.org is a tree counter, which shows over 35 billion trees planted out of 110 billion trees already scheduled to be planted.

https://theprint.in/world/pakistan-plans-to-plant-10-billion-trees-by-2023-to-restore-forest-cover/569129/

https://www.cbsnews.com/news/climate-change-india-plants-220-million-trees-in-a-single-day-to-save-the-planet/

https://www.abc.net.au/news/2019-10-17/green-walls-in-china-and-africa-keeping-deserts-at-bay/11602796 80 billion trees planned!

https://www.aljazeera.com/videos/2020/1/23/world-economic-forum-leaders-pledge-to-plant-1-trillion-trees

Chapter 9 Questions:

1. Why would China, India, and Pakistan plant billions of trees?
2. Why did the World Economic Forum say we need to plant one trillion trees?
3. Why is planting trees the only solution for worldwide carbon dioxide?

4. Web search wall of trees in Africa. See why trees release terpenes, which induce rain.

Chapter 10

Atmospheric Carbon Dioxide Doesn’t Freeze in the Mesosphere

The coldest place in the earth's atmosphere is the mesosphere. Pressure in the mesosphere is 1mb (1 millibar). At 1 millibar the freezing point of atmospheric carbon dioxide is –101 C. The temperature in the mesosphere is –90C. This is shown in the left and right charts in figure 10-1. Therefore, atmospheric carbon dioxide does not freeze in the atmosphere. This chapter is not the place to learn how a triple point graph works. Since one can learn this in any high school or college chemistry class, the authors will not explain it here.

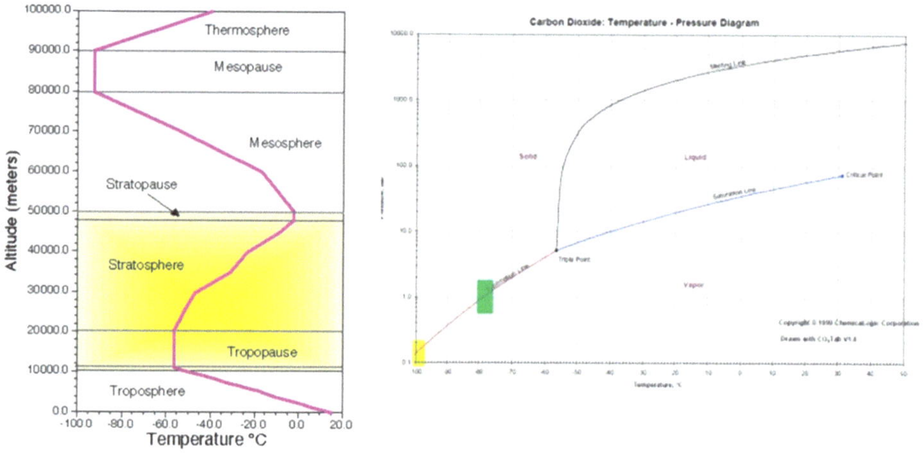

Figure 10-1 Temperature (left side) in mesosphere is –90C. Freezing point (right side) is –101C

Chapter 10 Questions:

1. Why does atmospheric carbon dioxide not freeze in the atmosphere?
2. What is the freezing point of atmospheric carbon dioxide at 1 millibar pressure?

Chapter 11

NIST and A Photosynthesis Experiment

Let's look at the short definition of the Scientific Method. The short definition: The scientific method begins with a hypothesis. If lots of data support the hypothesis, it may become a theory. If even more data support and verify the theory, the theory will become a scientific law.

The State of Oregon sanctioned an experiment in a non-treed area and a treed area. The treed area was 28 ppm lower carbon dioxide concentration than the non-treed area! Even in snowy December the difference was 15 ppm.

The carbon dioxide sensors used in this experiment were certified by the National Institute of Science and Technology. The data from the sensors are guaranteed by the US government. Carbon dioxide sensors were calibrated within 1 ppm of each other.

Scientific Law achieved! Hypothesis-> Theory with more than 4.5 million data points. Theory-> Scientific law with more than 4.5 million more data points.

The non-treed area sensor read consistently 394-494 ppm for 46 degrees north. The NOAA graph predicts 416 ppm! This is far outside that prediction. Figure 11-1 shows the non-treed area where HWY 217 merges onto HWY US 26 E. Figure 11-2 is the treed area at the Zoo exit. It is said we can't plant our way to lower atmospheric carbon dioxide. On the contrary, planting trees is the only way. See Chapter 1. The equilibrium for carbon dioxide in the atmosphere is 8.6 gt/yr. (8.6 billion tons per year) The spring in 2021 was drier than normal. This is what caused the photosynthesis to decrease in figure 11-3 in the spring.

Experimental Facts: US26E has 160k vehicles per day, according to the ODOT vehicle counter.

Photosynthesis equation. $6CO_2 + 6H_2O + \lambda \rightarrow C^6H_{12}O_6 + 6O_2$

Figure 11-1 is the non-treed area where HWY 217 merges to US26E

Figure 11-2 the treed area at the Zoo exit.

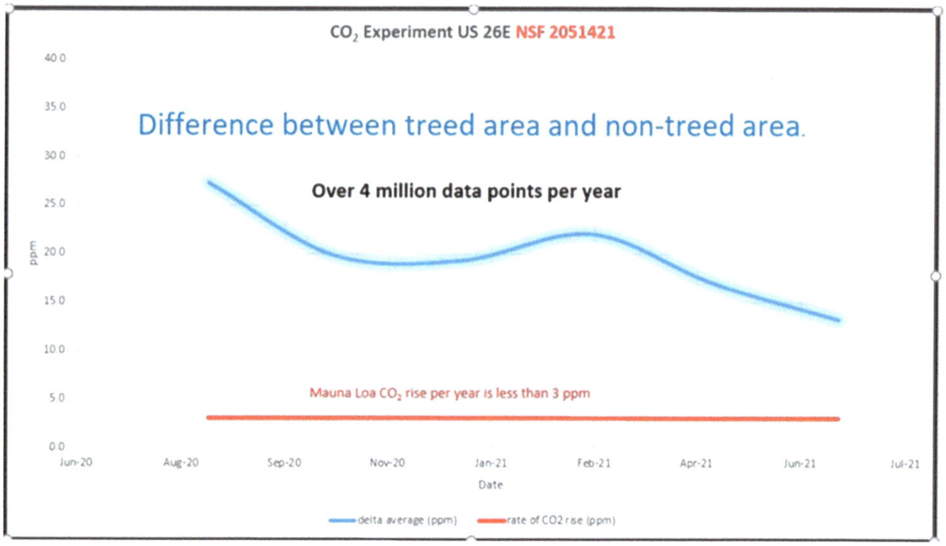

Figure 11-3 Graph of data from experiment.

On 9/23/22 we submitted a proposal (2246721) for funding from the National Science Foundation to replicate this experiment on the east coast of the USA, but the proposal was denied.

Chapter 11 Questions:

1. What is NIST?
2. Why in the spring of 2021 in western Oregon did the photosynthesis decrease?
3. The spring of 2022 in western Oregon was wet. Why did this increase photosynthesis?

Chapter 12

The Ocean is not a Sink for Atmospheric Carbon Dioxide

$$\text{Diffusion flux} = -D(dc/dx) \approx (-D(\Delta C/\Delta X))$$

Movement of carbon dioxide (not necessarily random) diffuses from high concentration to lower concentration. Example: diffusion of carbon dioxide is 2 cm/month toward the exosphere, not toward the ocean. The ocean is not a sink for carbon dioxide. Figure 12-1 shows carbon dioxide and all greenhouse gasses are well distributed by atmospheric winds. The D in the diffusion equation is the diffusion coefficient. A gas such as carbon dioxide is affected by pressure and temperature. The definition of pressure is 1 atmosphere at sea level. So, the pressure is not an effect worldwide. The temperature is warmer at the equator than the poles. Therefore, more diffusion flux would be present at the equator than the poles. Figure 15-2 is from the manuscript which triggered the junk science which said the ocean is a sink for atmospheric carbon dioxide. The recorded data do not reflect diffusion flux. Around 600 manuscripts claim that the ocean is a sink for carbon dioxide, but they only state that at the beginning of their manuscripts and never prove it.

Most scientists wrongly believe the ocean is a sink for carbon dioxide, which causes carbonic acid formation. Figure 12-3 shows that the only place on earth where ocean acidity is occurring is an area where human waste has been dumped by cruise ships since the 1950's. The decay of human waste consumes oxygen and produces carbon dioxide, which, in turn, changes to carbonic acid. Most news reports remove the Hawaii part of the graph. Figures 12-1, 12-4, and 12-5 show no acidity change.

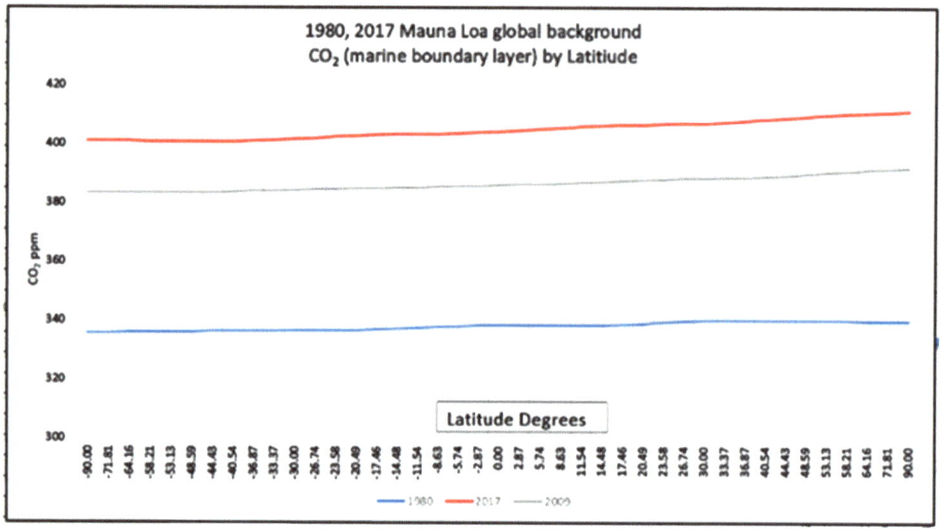

Figure 12-1 Almost equal distribution of greenhouse gases by latitude.

Mean rates of change in surface water pCO_2 and bulk water temperature of mixed layer (SST) estimated in six areas in the temperate South Pacific Ocean.

Locations	Lat.	Long.	Data period	Rate of change		No. of MOS.	Rate of change	
				pCO_2/SST (μatm y^{-1})	±σ (μatm y^{-1})		SST (°C y^{-1})	±σ (°C y^{-1})
Tahiti	15-20°S	135-145°W	11974-1997	2.00	0.39	16	0.033	0.034
Vanuata	20-25°S	165-175°E	1984-2006	1.30	0.27	35	-0.048	0.018
New Caledonia	25-30°S	170-180°W	1974-2005	1.05	0.09	17	-0.033	0.020
Tasmania	43-48°S	140-150°E	1984-2004	1.83	0.56	18	-0.051	0.062
New Zealand	45-50°S	170-180°E	1974-2006	1.42	0.30	37	0.001	0.027
S. of Tasmania	50-55°S	140-150°E	1984-2002	1.61	0.20	12	-0.044	0.064
Mean				1.5±0.3	0.30	-	-0.02±0.05	0.04

The rates are computed using mean monthly values. The temperatures are measured concurrently with pCO_2.

Figure 12-2 Data from PMEL manuscript. Not diffusion flux.

Hawaii Carbon Dioxide Time-Series

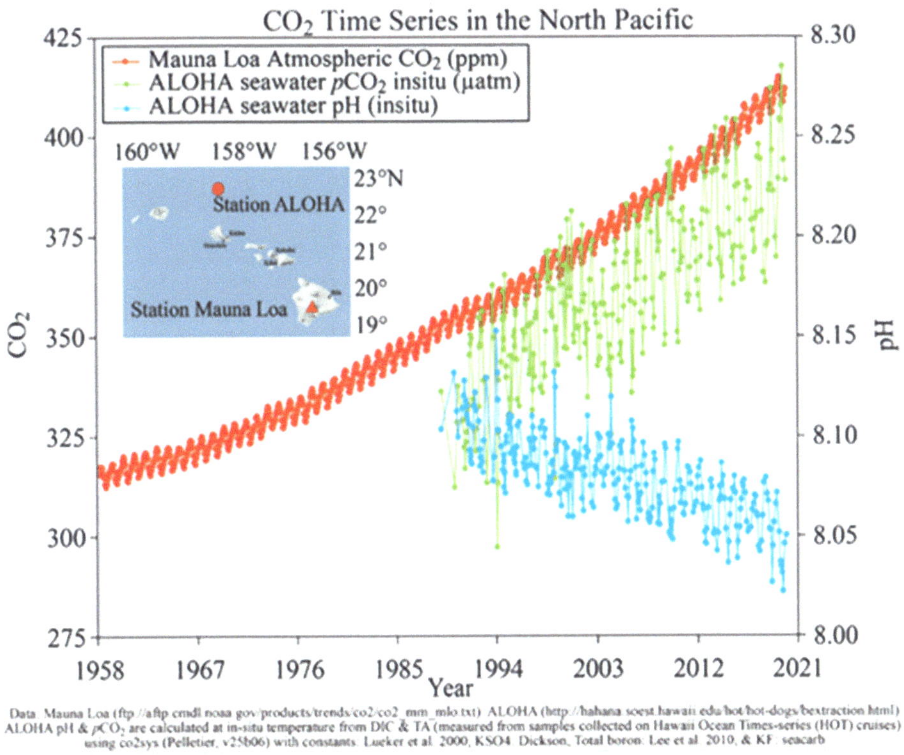

Figure 12-3 Hawaii is the only place on earth which shows an acidity change.

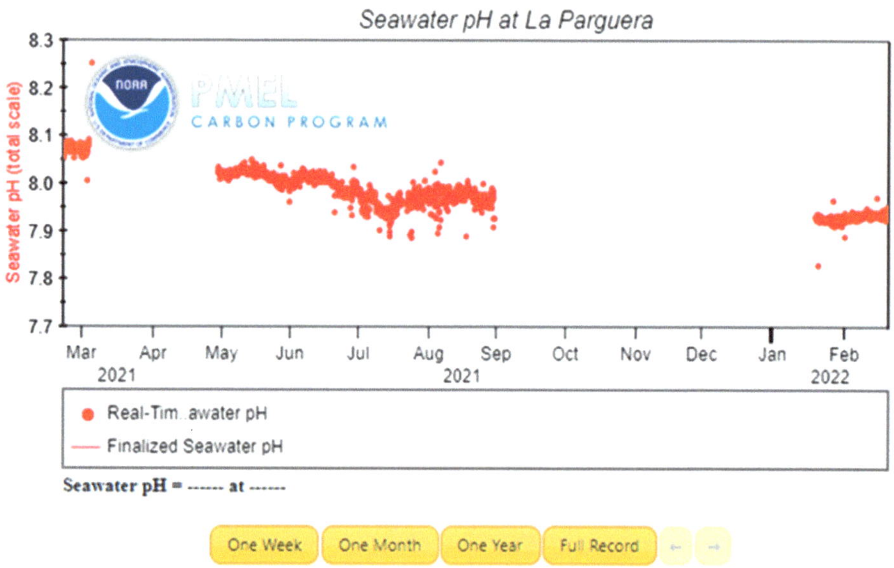

Figure 12-4

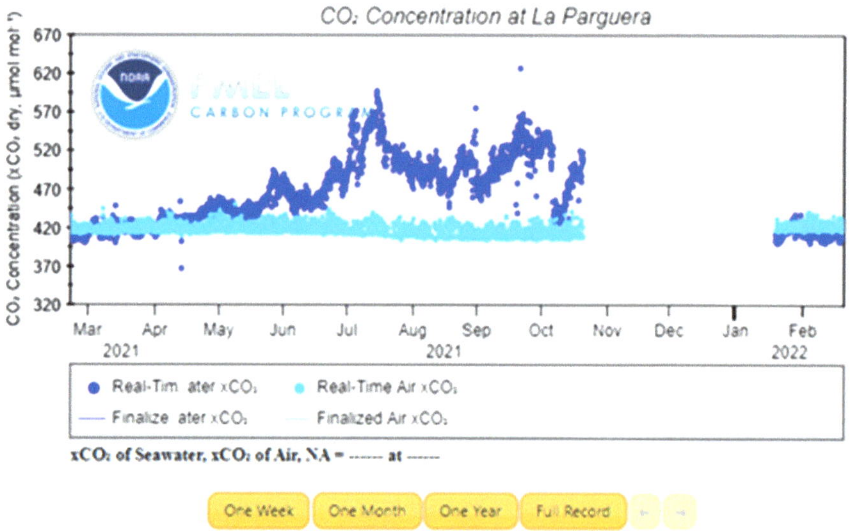

Figure 12-5

The spreadsheet has all data for the past 50 years. There exists no correlation between tropospheric carbon dioxide and ocean carbon dioxide. Rxy=0.65

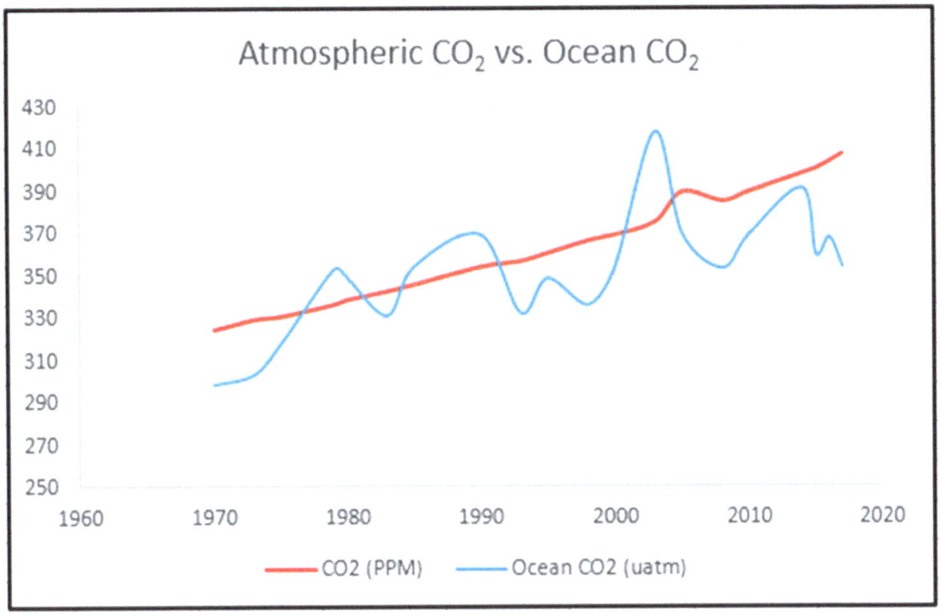

Figure 12-6 Atmospheric carbon dioxide vs. ocean carbon dioxide.

References:

12-1 La Parguera Mooring (17.95°N, 67.05°W) No change in CO2 or ph.

https://www.pmel.noaa.gov/co2/story/OA+Observations+and+Data

12-2 **Ocean Acidity**

Ocean acidity (or lack thereof. Tony Heller shows how the ocean acidity is the same as it's always been in this video. Ocean stupidifcation https://www.youtube.com/watch?v=ylpDngGm5cQ

Tony Heller's slides.

Figure 12.8 is from Wikipedia's false definition of ocean acidification. Note the bolded words. Estimated is not a scientific word and no estimation without lots of data is ever correct. Earth System Models project. Any model which is not benchmarked with data is a false model. Could undermine is not a scientific statement.

> *Ocean acidification is the ongoing decrease in the pH of the Earth's oceans, caused by the uptake of carbon dioxide (CO 2) from the atmosphere.[2] Seawater is slightly basic (meaning pH > 7), and ocean acidification involves a shift towards pH-neutral conditions rather than a transition to acidic conditions (pH < 7).[3] An estimated 30–40% of the carbon dioxide from human activity released into the atmosphere dissolves into oceans, rivers and lakes.[4][5] To achieve chemical equilibrium, some of it reacts with the water to form carbonic acid. Some of the resulting carbonic acid molecules dissociate into a bicarbonate ion and a hydrogen ion, thus increasing ocean acidity (H+ ion concentration). Between 1751 and 1996, surface ocean pH is* ***estimated*** *to have decreased from approximately 8.25 to 8.14,[6] representing an increase of almost 30% in H+ ion concentration in the world's oceans.[7][8]* ***Earth System Models project*** *that, by around 2008, ocean acidity exceeded historical analogues[9] and, in combination with other ocean biogeochemical changes,* ***could*** *undermine the functioning of marine ecosystems and disrupt the provision of many goods and services associated with the ocean beginning as early as 2100.[10]*

Figure 12.8

Figure 12.9 is Actual measured data in Monterey Bay California shows the pH range from 7.7 to 8.1. The range is much larger than the change is the Wikipedia definition. This is nine years of daily data collection.

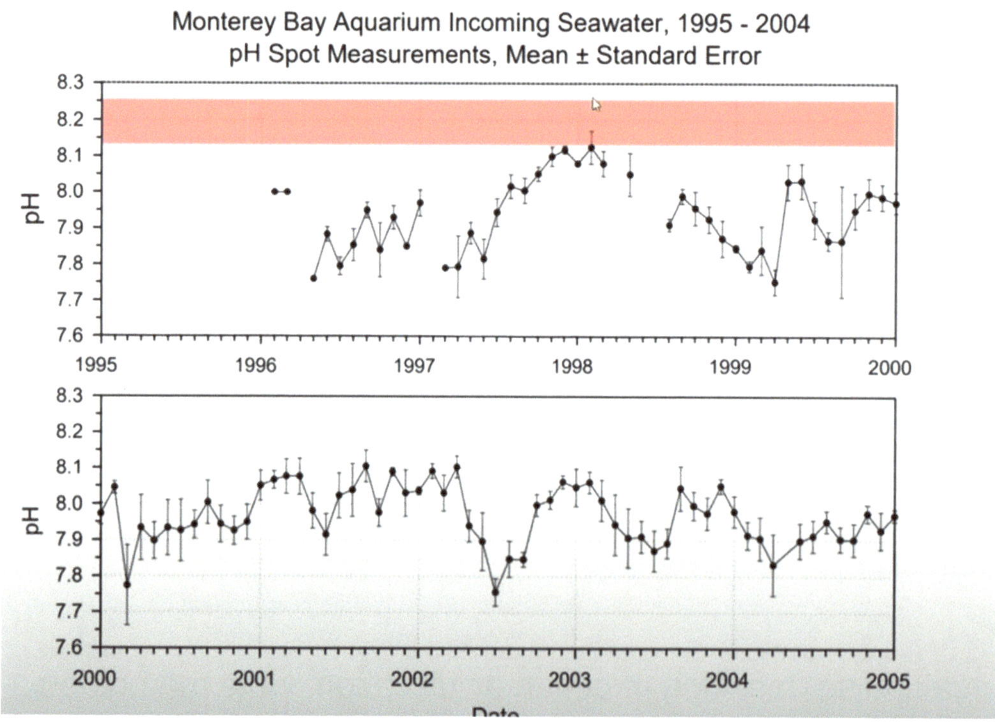

Figure 12.9

Figure 12.10.

Figure 12.10 is when life exploded on earth the carbon dioxide levels were twenty times more than they are now.

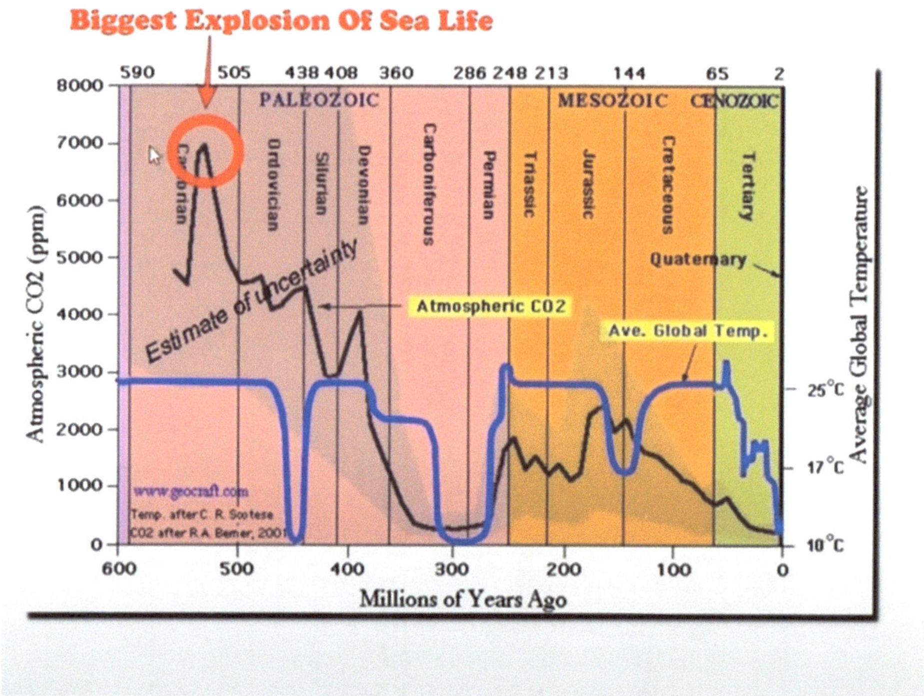

Figure 12.11

The ocean bottom is covered with Basalt rock from Volcanic eruptions. Basalt is very basic in ph. We could dump tons of acid into the oceans and the Basalt would neutralize it. The ph of the Oceans is controlled by the Basalt rock.

Figure 12.12

Akumal Mexico snorkeling.

Figure 12.13

Some of the reef looked like this. However, some areas a short distance away were dead. If Ocean Acidification we happening then all areas would be dead.

Chapter 13

The Intergovernmental Panel On Climate Change (IPCC) Reports Are Deliberate Falsifications

IPCC

Intergovernmental Panel on Climate Change

A Preliminary Experiment

On the Effect of Photosynthesis on Carbon Emissions

Applied Science: Nations Are Already Winning the Battle Against Climate Change

Sea Level Rise as Proof of Global Warming Debunked

So What's the Problem?

Mauna Loa Measurement Fraud

IPCC Watch Dog Group Calls Foul

The Futility of Reducing Carbon Dioxide Emissions As a Strategy for lowering Atmospheric CO2

Practical Applications for Individuals and Governments

Bio

EndNotes

ABSTRACT: Climate change hysteria is driven by the UN's Intergovernmental Panel on Climate Change (IPCC) Reports that are inaccurate and falsely skewing climate data. Moreover, the IPCC ignores properly benchmarked data resulting in invalid simulation results and faulty recommendations. This is the conclusion of a team of 35 doctoral level scientists who function as an official watchdog group to investigate, challenge and expose the duplicitous science on which the climate change agenda is perpetuated. It all started when Author, Dave White, presented his statistical analysis of climate change data to Dr. Mike Kuperberg (mkuperberg@globalchange.gov; mkuperberg@usgcrp.gov) of the National Academy of Sciences. Kuperberg is the Executive Director of the U.S. Global Change Research Program (USGCRP). He recognized the validity of the analysis and immediately submitted it to other scientists in his office. Their recommendation was for White to assemble a team of scientists to participate in the annual "Expert and Government Review (EAGR)" program of the Intergovernmental Panel on Climate Change (IPCC) reports.

The ranks of this team of doctoral level scientists soon swelled to over thirty, who collaborated in preparation of a college textbook on environmental science authored by White. Together they participated in the "EAGR" program and began exposing the junk science in the IPCC reports. In addition, they pressured the Nature Climate Change to replace Adam Yeeley, PhD PolySci, their chief editor until June 2020, whose scientific credentials were sorely lacking. Yeeley was allowing IPCC scientists to publish loosely referenced manuscripts and circular reference them in their reports.

This, of course, is a far cry from acceptable scientific procedure, resulting in IPCC reports that amount to deliberate science fiction. The IPCC writers self-identify as climate experts and inform journalists and governments worldwide in their reports on what to believe about climate change. Misleading reports result in government policies that negatively impact every person and business around the globe through unnecessarily harmful economic restrictions and taxation.

This team, comprised of over 35 doctoral-level scientists, most of them college professors, swiftly grew under Whites direction. Together, we collaborated on the preparation of a groundbreaking college textbook for environmental science in College, bearing my name. The involvement of our "Doubting Thomas" group in the "EAGR" program has become a platform to identify and address inherent problems that creep into the Reports. This collaborative effort with the ICPP is contributing significantly to the ongoing discourse on climate science within the scientific community (1, 2, 3,4, 6).

We also reviewed the NOAA 4 year report and it was garbage also (5).

Somehow the conclusions of the United Nations Intergovernmental Panel on Climate Change never made sense to me. As a Chemical Engineer and Statistician I sensed that the numbers just didn't add up. In spite of trillions spent on controlling emissions of carbon dioxide, why had virtually nothing been accomplished to lower atmospheric carbon dioxide?

In 2016 I decided to crunch some numbers and see what was actually going on. Step one was gathering all the pertinent data I could get my hands on. That included of climate change data of the National Academy of Sciences.

A Preliminary Experiment

On the Effect of Photosynthesis on Carbon Emissions

Based on years of experience, my statistical analyses demonstrated conclusively that carbon dioxide emissions are not the cause of the rise in carbon dioxide levels. If not, then what is? I reasoned that atmospheric carbon dioxide is a binary system statistically. The two causal factors are carbon dioxide emissions and loss of photosynthesis. Since carbon dioxide emissions are clearly not the cause, then there must exist a loss of photosynthesis cause and solution. This of course made perfect sense because many nations - Brazil in particular - had engaged in An orgy of deforestation for over 50 years. I wondered if reforestation along arterial highways serve the dual purpose of neutralizing carbon emissions and replenishing oxygen in the atmosphere?

I decided on a one-year experiment to test this hypothesis. This involved setting up censors to

measure carbon dioxide levels on Highway 26 as it winds up out of Portland headed toward the coast. For about a mile, the highway is flanked on both sides by a heavy growth of Douglas Fir. This experiment proved that the native trees and shrubs absorbed all the exhaust fumes of 160,000 cars and trucks on the highway each day. The findings of this experiment have profound implications for climate change policy at every level of government. Moreover, it demonstrates the relative ease by which the climate crisis may be resolved and the important steps already underway to resolve it by collaboration with some of the world's biggest nations.

Applied Science: Nations Are Already Winning the Battle Against Climate Change

China, India, Pakistan, and Peru are leading the way toward resolving the Carbon Dioxide imbalance in the

atmosphere. I called into the administrative departments of all of these governments,

presented the research and persuaded all of them to start planting trees. It turns out the

solution is hiding in plain sight. India stopped deforestation and is planting trees! China is

planting billions of trees! Pakistan planted 1 billion trees in 2018, 2 billion more in 2019, and

they will plant 8 billion more in the next four years! Peru stopped deforestation in 2020! And

now at last Brazil has turned the corner also with its new President as of 2023.

So what have been the results in terms of CO2 levels in the atmosphere. Although the IPCC is loathe to admit it, Carbon Dioxide levels have already stabilized. As demonstrated later in this article it is impossible for emissions control to have any such positive effect in so short a time due to the drag of Carbon Dioxide Residence Time. The only possible cause is increased photosynthesis.

Already planting 41 billion trees and the global garden greening atmospheric Carbon Dioxide minimum on October 4th was 407.51 ppm. Dr Pieter Tans said it should be 408.6+/- 0.5. For November the rise was -0.45 ppm. (11/1= 411.02, 4/20=410.57), November of 2017 it was 2.7 ppm rise. November 2018 1.85 ppm rise. 8 billion more trees scheduled in the next 4 years. We can easily plant 100 billion trees in the USA and in 10 years will consume an extra 10 billion tons annually. The aforementioned experiment went from Hypothesis to Scientific law we can plant native shrubs and trees next to roads and in just ten years they will consume all the carbon dioxide from the vehicles. This is applicable +/- 50 degrees from the equator.

Sea Level Rise as Proof of Global Warming Debunked

So What's the Problem?

The IPCC points to an alleged rise in sea level and glacier melting as proof of global warming and the need for more draconian throttling of the world economy. They rely on the Jason-3 satellite measurements, which have a minimum resolution of 25mm, to report a 3mm rise per year by measuring a location every 10 days. However, this measurement is meaningless because the reliability of any measurement below minimum resolution, drops exponentially below 50% of the minimum resolution, which is reported in the document review for WG I AR6 **(1).**

Thus, the only accurate measure of sea level are the tide gauges and they show almost no sea level change. The Views of Three Sea Level Specialists, Mörner NA, Wysmuller T and Parker confirm this finding (2). Moreover, the Jakobshavn Glacier in Greenland has actually grown for the fifth year in a row, as of **2023.** This is the big one that Al Gore and others have led us to believe would melt and cause the oceans to rise 15 feet. A documentary called Climate Hustle II came out in October 2020 to expose this. In addition, if sea levels are rising we would expect acidity in the oceans to be diluted. But there's no evidence that this is occurring. Tony Heller shows how the ocean acidity is the same as it's always been in the video, Ocean Stupidification.

With global temperatures and sea levels stable, the threat of a climate emergency is clearly way overblown. Nonetheless, national deforestation policies over the past 50-75 years have created an imbalance that needs to be corrected. The only worldwide manuscript for NetZeroCO2E (3).

Mauna Loa Measurement Fraud

The Mauna Loa volcanic cone on Oahu's west coast was selected back in 1980 as an ideal spot to collect atmospheric readings. This is because of its height and its relative isolation from environmental contamination due to the ocean on its western slope. These readings have therefore been deemed representative of worldwide CO2 levels. The oscillation at Mauna Loa starts as a very strong signal in South America and then fans out larger and larger until Barrow's Alaska. The countries in South America burn the Amazon Rainforest, the densest forest in the world, from October/ November through May of the next year. Since 1950, an average of 30 million acres per year have been deforested and burned. So much Carbon Dioxide has been released that the trees and plants have grown too fast and died. This massive decay is what caused the Amazon Rainforest to switch to an oxygen sink and carbon dioxide producer. Hundreds of papers have been published on this. Currently, the Amazon output of carbon dioxide is 15 GTyr-1 of CO2. With 41 billion trees planted on cctruth.org which are consuming carbon dioxide and lowering carbon dioxide to 330 ppm by 2031 NOAA is committing 19 ppm of Fraud (4). The scientists who committed this fraud have been fired. The Amazon Rainforest deforestation is a 0.98 cause and effect to the rise of carbon dioxide since 1957.

Amazon Rainforest Rxy =-0.99 The loss of oxygen worldwide is a 0.99 cause and effect to the destruction of 2 billion acres of the Amazon Rainforest since 1950! The correct solution is to stop non-sustainable deforestation of those forests like the Indian and Amazon Rainforests and plant 200 billion native trees and shrubs.

I registered a complaint with the Department of Commence Inspector general about Mauna Loa CO2 fraud. They started investigating on 4/24/20. We encourage anyone to download the rain-forest stop document and follow it weekly.

IPCC Watch Dog Group Calls Foul

The time had come to start pushing back against the United Nations IPCC. In 2019 I presented my statistical analysis of climate change data to Dr. Mike Kuperberg (mkuperberg@globalchange.gov; mkuperberg@usgcrp.gov) of the National Academy of Sciences. Kuperberg is the Executive Director of the U.S. Global Change Research Program (USGCRP). He recognized the validity of the analysis and immediately submitted it to other scientists in his office. Their recommendation was for White to assemble a team of scientists to participate in the annual "Expert and Government Review (EAGR)" program of the Intergovernmental Panel on Climate Change (IPCC) reports as a Fact Checker. The ranks of this team of doctoral level scientists soon swelled to over thirty, who collaborated in preparation of a college textbook on environmental science authored by White. Together they participated in the "EAGR" program and began exposing the problems in the IPCC reports. In addition, they pressured the IPCC to replace Adam Yeeley, PhD PolySci, the chief editor of Nature Climate Change, whose scientific credentials were sorely lacking. Yeeley was allowing IPCC scientists to publish loosely referenced manuscripts and circular reference them in their reports.

This, of course, is a far cry from acceptable scientific protocol, resulting in severe distortion of the IPCC reports. Whether this is negligence or simply ineptitude we can only speculate.

Either way the IPCC writers self-identify as climate experts and inform journalists and governments worldwide in their reports on what to believe about climate change. Misleading reports result in government policies that negatively impact every person and business around the globe through unnecessarily harmful economic restrictions and taxation. This of course is the reason our watchdog group was activated – to assist IPCC in its daunting task.

In our initial PhD review of IPCC working Group 1 (ASR1.5), in the first order draft for Ar6 we identified and exposed the flaws in the global warming potential model (5). This model assumes equal greenhouse gas (GHG) concentrations in the atmosphere, which is illogical. Carbon Dioxide, for example, is more than 200 times the concentration of methane. Furthermore, we found in

Annex 2, a table with the correct order of GHG effects. Annex2 had results from measurement of Greenhouse gas effect which is similar to the Rutgers University benchmark data. Water vapor is 88% effect, CO_2 is 9% effect, Methane is 0.3% effect. The IPCC was ignoring their own data! Any such model which fails to benchmark against these established effects is by definition an inadequate model. At least 30 members of our review team sent a personally endorsed copy of our review, asking IPCC to correct their inaccuracies. (5) The result? The scientific review of 30 climate change experts, virtually all of them college professors or research specialists, was ignored. That makes the AR6, Working Group one report misleading, and counterproductive as a whole. A house built on sand cannot withstand the storm. However, in this case rather than correct their faulty model, IPCC editors simply deleted the Greenhouse Gas table from Annex 2 in their final draft for AR6. Instead of making the necessary changes to their model they deleted the damning benchmark data in Annex 2. Sadly, this illustrates how unreliable and apparently agenda-driven the IPCC has become. It is literally garbage in, garbage out. You cannot have an accurate model without accurate benchmark data to validate it. The New Book Unsettled shows this and is correct. The outcome is predictable: In this instance the Reports had been written to support a political agenda. These findings are documented at https://cctruth.org/ipcc.pdf with links to the Reports and spurious data.

But wait a minute. Everybody knows that 97% of the world's scientists agree with the IPCC that the earth is warming at an alarming rate, right? Let's take a look. The 97% consensus figure was derived from three hundred and thirty manuscripts published between 2009 and 2012 which were pro the false agenda that were cherry-picked for review by the IPCC. Excluded from the review and survey of scientists were more than seven hundred manuscripts written by scientists who presented statistics and conclusions different from those desired. That reduces agreement from 97% to 33%. If IPCC fails this basic test of statistical sampling, how can we rely on what they are telling us about climate change?

The Futility of Reducing Carbon Emissions

As a Strategy for lowering Atmospheric CO_2

The primary reason that controlling emissions to lower atmospheric CO_2 will not work has to do with the phenomenon of Residence Time. Residence Time is how long a molecule will stay in a

location before being released. So the key question is how long will current levels of Carbon Dioxide

remain in the atmosphere, all other things being equal?

To answer this question, I sent out a survey email to 4000 climate change scientists

relating to atmospheric Carbon Dioxide residence time. Most scientists who responded said it was 200-400

years. However, one scientist sent me his meta research on published papers, which show

residence time for Carbon Dioxide ranging from 150 years to 700 years.

CO2 Residence Time (Years)	Author	Year
700	Allen	2009
610	Zickfeld	2013
500	Matthews	2008
300	Plattner	2008
270	Cao	2010
230	Zickfeld	2012
220	Solomon	2012
220	Knutti	2012
210	Gillett	2011
180	Frolicher	2010
150	Hare	2006

Even the IPCC admits as much, although they leave a big margin of error. A 2003 IPCC report shows residence time increased from between 5 and 200 years.

In the Global Carbon Dioxide rise there is still no effect from these worldwide happenings which would have lowered carbon dioxide by up to 90% for almost two years. Oil embargo in the 1970's, for almost two years the worldwide carbon dioxide emissions would have dropped by 90% Multiple recessions each one the worldwide carbon dioxide emissions would have decreased by 40% for at least one year(6).Worldwide recession in 2009. A 70% reduction in emissions of carbon dioxide for almost two years. COVID-19 pandemic. A 6% reduction in emissions for 1.5 years. You can clearly see no signature from these events in the NOAA data. In 2016, I queried Dr. Jim Hansen and two other prominent climate-change scientists about current emissions and their effect on atmospheric Carbon dioxide. They replied that emissions had been flat since 2014, and that not only was atmospheric CO_2 still increasing but the rate of rise was still increasing. I asked them how this could be happening--if emissions were the cause of atmospheric CO_2 increase? They responded that we must wait another 470 years for anything we do with emissions to show an effect. In other words, anything we do with CO_2 emissions has not and will not have any effect on atmospheric CO_2 for hundreds of years.

That's because the residence time for atmospheric carbon dioxide is no less than 150 years. This is why everything we have done so far to lower emissions of CO_2 has had absolutely no effect on the atmospheric CO_2 rise. It's like starting at the beginning of a 100-meter foot race, when your opponent is starting 1 meter away from the finish line.

So even assuming a reduction in 45% of fossil fuel emissions by 2030 the decrease of carbon emissions will be offset by increases in population. Thus, the slope for increase of atmospheric CO_2 remains flat; i.e., not increasing. However, rate of rise is increasing, with the current rate at almost 3 ppm increase per year. Then assume we run out of oil in 100 years, so CO_2 emissions drop by another 55% to zero. Atmospheric CO_2 lowers to a minimum at year 2650 and then increases. We will never reach equilibrium. This is because we have massive loss of photosynthesis. A decrease of photosynthesis got us into this mess and only an increase of photosynthesis will get us out of it. We have shown that atmospheric CO_2 is still rising even faster although the CO_2 emissions rise has slowed by 50%. We showed how cap and trade policies would have zero effect on the rise of atmospheric carbon dioxide because the equilibrium point is too low. The strategy with the most positive effect on lowering atmospheric CO_2 is increasing photosynthesis.

This will in turn increase the equilibrium point to over 100 GTyr-1.

The only way to lower atmospheric carbon dioxide is to increase photosynthesis. The correct and only solution is to stop non-sustainable deforestation and burning of large rainforests (such as those in India and the Amazon River Basin and its tributaries) and to plant 200 billion native trees and shrubs, especially in those areas that have been deforested. This will cause atmospheric CO_2 to lower to 330 ppm by 2031 (see Graph 11). Use ecosia.org for internet search engine (7) it plants trees.

Dr. Tom Crowther (Crowther et al. 2019) published a paper on increasing photosynthesis with recommendations of where to plant. The study found that most of the land suitable for restoring forests is in six countries: Russia (151 million hectares), USA (103 million hectares), Canada (78 million), Australia (58 million), Brazil (50 million), and China (40 million). Appendix 1 shows how to plant the trees and shrubs. This will increase consumption of CO_2 to over 100 Gt yr-1. Not every forest hectare is equivalent in photosynthesis consumption of CO_2. Rainforests consume 90-100 tons per hectare of carbon dioxide per annum. Other forests are from one quarter to 8 tons per hectare per annum.

Practical Applications for Individuals and Governments

We have learned that focusing on increasing photosynthesis is far easier, quicker and vastly less

expensive than futile attempts to decrease vehicle and industrial emissions, which take

hundreds of years to have any effect. We've already seen dramatic results at decreasing

atmospheric CO_2 in months as opposed to waiting hundreds of years for the emissions based

approach to take effect. So here's what we can do by working together as individuals and

governments to lower atmospheric Carbon dioxide quickly.

1. Put pressure on Brazil and other Amazon rain-forest countries to stop deforestation ASAP. Also stop the biomass burning that puts 300 million tons of carbon dioxide into the atmosphere each year. This has caused 50ppm of the recent rise in atmospheric carbon dioxide concentration. Then after 10 years finish burning what is needed at 10% per year for 10 years.

2. Provide space where public can come and plant trees and shrubs. All government-owned lands. Very small cost. Need website with document for each planting area.

3. Plant shrubs in all freeway medians and sides. This is revenue plus in a two-year cycle. Plant native shrubs at a minimal spacing so all light is used in photosynthesis. This will take in 1 ton of CO2 emissions per acre per year right at the source. The space would not need to be mowed every week in the summer.

4. Get schools involved and planting massive number of trees and shrubs. In their property and the government | property as in 1 above.

5. Parks can add trees and shrubs.

6. Close any climate change research group. Not needed, unless doing photosynthesis work.

7. Tax incentive for business to plant trees and shrubs.

8. Wild fire attention. Get a retainer for the 747 plane and use it from the start on any wild fire.

9. Forest management by “strip logging” which was developed by Oregon State Forestry. This strip 30 to 60 yards wide

(depending on the height of the trees) will provide ongoing logging opportunities, making these cuts. The side trees and

shrubs will naturally reseed these cuts. These seeds are matched genetically to the local soil and climate. They grow

much faster because of this. No reseeding is needed or desired. These cuts make an excellent firebreak.

10. This drone can plant 40,000 trees per day.

Bio: Dave White Is a Chemical Engineer and Statistician. In addition to his research he teaches at HyMarkAcademy.com. With a lifetime of experience in research science, Dave is far more than just another science teacher with a degree. His research is having an international impact. He's currently working on exposing the myths surrounding Climate Change and implementing real solutions. He has 30 years' experience since graduation in 1984, promoting environmental responsibility and survival of all species. This wealth of practical experience enriches all of his classes and engages his students in real science.

References

1. https://cctruth.org/WGII_SOD_Ar6.xlsx Working Group II Second Order Draft for Ar6 uploaded by team members.
2. https://cctruth.org/comments_ar6wg1_sod.xlsx Working Group I Second Order Draft for Ar6 uploaded by team members.
3. https://cctruth.org/comments_ar6wg2_sod.xlsx Working Group II Second Order Draft for Ar6 uploaded by team members (updated prior to final)
4. https://cctruth.org/ comments_ar6wg3_fod.xlsx Working Group III (mitigation) Second Order Draft for Ar6 uploaded by team members.
5. https://cctruth.oorg/comments_srocc_sod.xlsx NOAA report every 4 years uploaded by team members.

6. https://cctruth.org/Comments_WGIAR6.xlsx Working Group I Second Order Draft for Ar6 uploaded by team members.
7. Ecosia.org

++++++++++End of Report++++++++++

Addendum not part of article but more proof.

In our initial PhD review of IPCC working Group 1, in the first order draft for Ar6 we identified and exposed their inaccurate global warming potential model. This model assumes equal greenhouse gas (GHG) concentrations in the atmosphere, which is absurd. Carbon Dioxide, for example, is more than 200 times the concentration of methane. Furthermore, we found in Annex 2, a table with the correct order of GHG effects. Annex2 had results from measurement of Greenhouse gas effect which is similar to the Rutgers University benchmark data. Water vapor is 88% effect, CO2 is 9% effect, Methane is 0.3% effect.

The IPCC was ignoring their own data! Any such model which fails to benchmark against these established effects is an inaccurate model. At least 30 members of our review team sent a personally endorsed copy of our review, asking IPCC to correct their inaccuracies. The result? The IPCC working group one deleted the data in the table in Annex One instead of deleting their junk model. The scientific review of 30 climate change experts, virtually all of them college professors or research specialists, was ignored.

That leaves the AR6, report bogus, misleading, and counterproductive as a whole. A house built on sand cannot withstand the storm. However, rather than correct their error, IPCC editors simply deleted the Greenhouse Gas table from Annex 2 in their final draft for AR6. Instead of making the necessary changes to their model they deleted the damning benchmark data in Annex 2. This is how unscrupulous and agenda-driven the IPCC has become. It is literally garbage in, garbage out. You cannot have an accurate model without accurate benchmark data to validate it. Then published in a journal whose former chief editor held a PhD in Political Science with very little acumen in the hard sciences. The outcome is predictable the Reports have been written to support a political agenda. These findings are documented at https://cctruth.org/ipcc.pdf with links to the Reports and spurious data.

IPCC Reports

Disclaimer: Sometimes the IPCC makes changes without notification. For example, the Executive Summary of the Mitigation Chapter added our review paragraph. However now to confuse people they start out every paragraph with the same words. Previously this was not done. Also they changed the numbering scheme for the chapters. The difference is they are now beginning four paragraphs with this statement, "**Limiting warming to 1.5°C depends on greenhouse gas (GHG) emissions**". The three paragraphs that start with this statement have nothing to do with our review and are just there to mislead people. In fact, they still state inaccuracies they've been told about on several occasions such as methane gas is the worst greenhouse gas. However, by scientific measurement, it is clear that methane gas is 0.29% effect and water vapor is 89.4% greenhouse gas effect. Second Order Draft (SOD) for AR6 review (5).

[Dave, this paragraph (above) is not clear] [In the paragraph below we claim 23, not 30]

The first review for SR 1.5 was 23 PhD's. Ar6 was 30 PhD's.

To review, our 23 PhD review of IPCC working Group 1 first order draft for Ar6 identified the fatal errors in the IPCC global warming potential model. This model assumes equal greenhouse gas (GHG) concentrations. Such an equal concentration never occurs in the real world. For example, carbon dioxide is more than 200 times the concentration of methane. Surprisingly, in Group 1, we found in Annex 2, a table with the correct order of GHG effects, which was being ignored. Any model which fails to benchmark the correct data and the correct order is a fraud. Our review was sent by at least 23 team members to inform IPCC of their obligation to benchmark their Annex 2 table to the correct order of GHG effects.

Unfortunately, IPCC chose not to benchmark their final draft of Ar6, but instead chose to delete the incriminating table in Annex 2. This left their sham GWP model intact. This wasn't just a simple oversight of the benchmarked data. The IPCC purposely concealed the fact that their alleged scientific model was fallacious. This elevates the concept of "cooking the data" to a whole new level.

How does the review process work? Typically, at least twenty-three doctoral level scientists participate in the "Expert and Government review" program for the IPCC reports. We find a vast array of problems in them. Each member of the team downloads the reports by various "working groups" of the IPCC. They examine each report line by line and then meet online to decide what changes to submit. Then each member submits the same report to the IPCC, a total of more than twenty-three submissions. The total PhD's from the IPCC for each chapter ranges from 15 to not more than 20. When 30 of us submit the same changes we mostly get those changes done like Working Group one.

For example, in the IPCC mitigation chapter, Jim Skea concluded that we need to lower atmospheric carbon dioxide emissions by 45% by 2030. However, the statement in the chapter he was basing that goal on was buried on page 95 and the equilibrium statements had no external references (citations) to any published manuscripts. They completely made it up! Also buried on page 101 was a statement that the probability of their solution to work is only 66%. After we submitted our review, they moved these items up into the 5th paragraph of their Executive Summary on page 6 for Ar6.

https://www.ipcc.ch/report/ar6/wg3/downloads/report/IPCC_AR6_WGIII_FullReport.pdf

This link is also on the college textbook page on cctruth.org

We performed an expert review of IPCC (Intergovernmental Panel on Climate Change) SR 1.5 Chapter three “Mitigation” .https://cctruth.org/expert_review_SR1.5_mitigation.pd f . These are the key findings: Their equilibrium statements had no references to any published manuscripts.

Yet not everything is ignored. One of the chapter scientists replied and said they are not equilibrium statements and that they are from simulations. I showed their simulations to a friend who has 27 years' simulation experience and he broke into uncontrollable laughter. Further down in their document was the only probability estimate they made of only 50-66% that their solution for lowering emissions will work. I sent this to around 1000 scientists, the worldwide media, the UN and IPCC scientists. The media ignored it, however, IPCC working Group 1 and 3 saw our expert review ability and invited us to review their reports for AR6.

https://cctruth.org/comments_ar6wg3_fod.xlsx is already accepted for WG 3.

2019 IPCC SR 1.5 Chapter 3 “**Limiting warming to 1.5°C depends on greenhouse gas (GHG) emissions over the next decades, where lower GHG emissions in 2030 lead to a higher chance of keeping peak warming to 1.5°C (*high confidence*). Available pathways that aim for no or limited (less than 0.1°C) overshoot of 1.5°C keep GHG emissions in 2030 to 25–30 $GtCO_2e\ yr^{-1}$ in 2030 (interquartile range). This contrasts with median estimates for current unconditional NDCs of 52–58 $GtCO_2e\ yr^{-1}$ in 2030** (https://www.ipcc.ch/sr15/chapter/chapter-2/, Page ES, 5th paragraph). Now their Executive Summary (https://cctruth.org/es.pdf) shows this statement with no references and their probability of 66%. I sent four emails asking them where these numbers came from. A research scholar at The International Institute for Applied Systems Analysis (IIASA) Schlossplatz 1, A-2361 Laxenburg, Austria replied:

"Dear Dave, Thank you very much for your question on the assessment of quantitative pathways in the SR15. The statement is taken from Table 2.4, bottom section, third row, first column, rounded to multiples of 5. The assessment in this table is based on the ensemble of quantitative pathways compiled by the IAMC and IIASA for the IPCC SR15 process

"(https://doi.org/10.22022/SR15/08-2018.15429). The Python script for preparing this table is available under an open-source license at https://data.ene.iiasa.ac.at/sr15_scenario_analysis/assessment/sr15_2.3.3_global_emissions_statistics.html (see https://doi.org/10.22022/SR15/08-2018.15428 for the scientific reference of the assessment notebooks).

"Neither the statement nor the table does make any assertion about an equilibrium; it is merely an assessment of the pathways at a specific point in time [bold added]. I do hope that this clarifies your request. The International Institute for Applied Systems Analysis (IIASA) Schlossplatz 1, A-2361 Laxenburg, Austria."

Please note! This faulty simulation has us reach equilibrium at 2050!

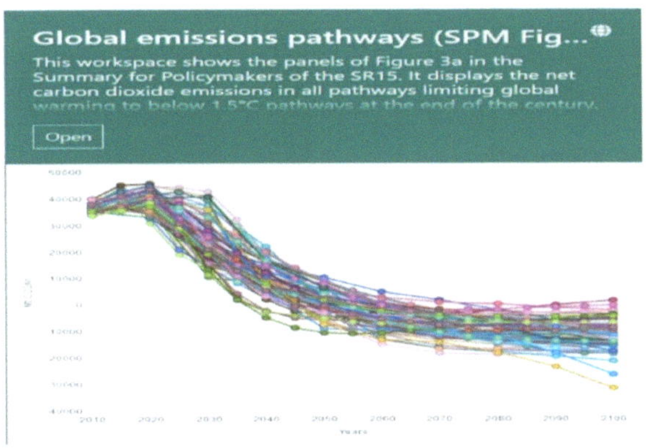

Their simulations are meaningless because they have no boundary conditions. Their simulation shows NetZero at zero in 2050. However, the IPCC and UN have propagated a fallacious 12-year doomsday scenario. Dr. Kevin Dayaratna testified at the Oregon Carbon group with the correct use of their simulations.

This is why nothing they have predicted has or will come to pass: https://ctruth.org/DAYARATNA.mp4

Earlier I sent this review to about 5000 scientists and major worldwide media by email. One NOAA scientist replied that I should go after the publishers of the IPCC "crappy" manuscripts. I thanked him and replied that I would if I had a large staff of scientists. I showed their simulations to an expert in simulations and he burst into uncontrollable laughter. Around December 15th 2019 I sent it to all other than Chapter two IPCC scientists. Our review was sent to the other 200 IPCC scientists who essentially agreed with the review we provided.

Rare Use of Probability

"For limiting global warming to below 2°C **with at least 66% probability** [bold added] CO_2 emissions are projected to decline by about 25% by 2030 in most pathways (10–30% interquartile range) and reach net zero around 2070 (2065–2080 interquartile range).1 {2.2, 2.3.3, 2.3.5, 2.5.3, Cross-Chapter Boxes 6 in Chapter 3 and 9 in Chapter 4, 4.3.7} (p 21.3, Table 2.1).

"No pathways were available that achieve a greater than **50-66% probability of limiting warming below 1.5° C** [bold added] during the entire 21st century based on the MAGICC model projections" For limiting global warming to below 2°C with at least 66% probability CO_2 emissions are projected to decline by about 25% by 2030 in most pathways (10–30% interquartile range) and reach net zero around 2070 (see p. ES, Paragraph 5).

TABLE 2.1

Classification of pathways that this chapter draws upon, along with the number of available pathways in each class

The definition of each class is based on probabilities derived from the MAGICC model in a setup identical to AR5 WGIII (Clarke et al., 2014), as detailed in Supplementary Material 2.SM.1.4.

PATHWAY GROUP	PATHWAY CLASS	PATHWAY SELECTION CRITERIA AND DESCRIPTION	NUMBER OF SCENARIOS	NUMBER OF SCENARIOS
1.5°C or 1.5°C-consistent**	Below-1.5°C	Pathways limiting peak warming to below 1.5°C during the entire 21st century with 50–66% likelihood*	9	90
	1.5°C-low-OS	Pathways limiting median warming to below 1.5°C in 2100 and with a 50–67% probability of temporarily overshooting that level earlier, generally implying less than 0.1°C higher peak warming than Below-1.5°C pathways	44	
	1.5°C-high-OS	Pathways limiting median warming to below 1.5°C in 2100 and with a greater than 67% probability of temporarily overshooting that level earlier, generally implying 0.1–0.4°C higher peak	37	

(Probability Table 2.1 page 21.3)

The probability is actually zero because the average residence time for atmospheric CO2 emissions is 150 years (IPCC 2003) **No business would spend such a significant amount of money (2.8 trillion dollars already spent worldwide) on a project with only a 50-66% chance of success.**

For ar6 which came out March 20th 2023 the mitigation chapter moved this to page 6 paragraph b.1.3. We forced them to do it https://www.ipcc.ch/report/ar6/wg3/downloads/report/IPCC_AR6_WGIII_Fu llReport.pdf You can see zero references in that paragraph. Only footnotes.

Citation

"This chapter should be cited as: Rogelj, J., D. Shindell,
K. Jiang, S. Fifita, P. Forster, V. Ginzburg, C. Handa, H. Kheshgi, S. Kobayashi, E. Kriegler, L. Mundaca, R.
Séférian, and M.V.Vilariño, 2018: Mitigation Pathways Compatible with 1.5°C in the Context of Sustainable
Development. In: Global Warming of 1.5°C. An IPCC Special Report on the impacts of global warming of 1.5°C above pre-industrial levels and related global greenhouse gas emission pathways, in the context of strengthening the global response to the threat of climate change, sustainable development, and efforts to eradicate poverty [Masson-Delmotte, V., P. Zhai, H.-O. Pörtner, D. Roberts, J. Skea, P.R. Shukla, A. Pirani, W.
Moufouma-Okia, C. Péan, R. Pidcock, S. Connors, J.B.R. Matthews, Y. Chen, X. Zhou, M.I. Gomis, E. Lonnoy, T. Maycock, M. Tignor, and T. Waterfield (eds.)]. In Press" (p. 93)

Use of Unscientific Terms

The document uses the unscientific terms *highly* (or otherwise) *likely* six times, *unlikely* three times, and *highly* (or otherwise) *confident* sixty-two times. In every case, percent probability must be used.
Planting Native trees is the only way to lower Atmospheric carbon dioxide to 330 ppm by 2031.

The IPCC follows a false agenda and a false GWP (Global Warming Potential) Calculation, neither of which is based on reality. Their GWP calculation assumes equal greenhouse gas concentrations of methane, nitrous oxide and carbon dioxide and other gases, which will never happen in reality. If we did have equal concentrations of N_2O (laughing gas) for instance, the people in the world would have silly smiles on their faces and high-pitched voices. IPCC Working group I, second order draft (SOD) Annex II found 14 published manuscripts which show the same data as Dr. Blasings. These were published prior to the GWP and the IPCC ignored them. We put this finding in our review for working group 1. They ignored it and deleted the 14 manuscripts! Any model which is not verified by data is a false model. The correct order of greenhouse gases CO2 then CH4 then N2O then NO (highest effect to lowest effect) Dr. TJ Blasing exposed the greenhouse gases with longwave radiation and was thus able to calculate the actual effect.
http://cctruth.org/index.php/ghg/ Methane is 0.5 watts/m^2. CO_2 is 1.94 watts/m^2. The media should not believe the IPCC or the UN when it comes to climate change. Dr. Hal Dorian passed away 4/28/20. His memorial. He is one of the NASA scientists who helped write our proposal. We dedicate our proposal to him.

Gas	Pre-1750 tropospheric concentration[1]	Recent tropospheric concentration[2,3]	GWP[4](100-yr time horizon)	Atmospheric lifetime[5](years)	Increased radiative forcing[6] (W/m^2)
Concentrations in parts per million (ppm)					
Carbon dioxide (CO_2)	~280[7]	399.5[2,8]	1	~ 100-300[5]	1.94
Concentrations in parts per billion (ppb)					
Methane (CH_4)	722[9]	1834[2]	28	12.4[5]	0.50
Nitrous oxide (N_2O)	270[10]	328[3]	265	121[5]	0.20
Tropospheric ozone (O_3)	237[1]	337[2]	n.a.[3]	hours-days	0.40

Planting trees is 100% probability to lower atmospheric carbon dioxide.

Residence Time of Atmospheric CO_2

Residence time is how long a molecule will stay in a location before being released. Like standing water in your kitchen, sink. The water is residing longer. A 2003 IPCC report shows residence time increased from 5 to 200 years. Dr. TJ Blasing shows 100-300 years. In 2016, I emailed Dr. Jim Hansen and two other prominent climate-change scientists that emissions had been flat since 2014, but that atmospheric CO_2 was still increasing and the rate of rise was still increasing. I asked them how this could be happening--if emissions were the cause of atmospheric CO_2 increase. **They said we must wait another 470 years for anything we do with emissions to show an effect.** Anything we do with CO_2 emissions has not and will not have any effect on atmospheric CO_2 for hundreds of years. However, the residence time for atmospheric carbon dioxide is 150 years. This is why everything we have done to lower emissions of CO_2 has had zero effect on the atmospheric CO_2 rise. https://cctruth.org/residence_time.pdf Below are the constraints I used. Even at average residence time of 100 years Mauna Loa never stays low.

Facts

Residence time was 5 years, Now more than 150 years. Recently I sent out a survey email to 400 climate change scientists about atmospheric CO2 residence time. Most scientists said 200-400 years. One scientist sent me his research of published papers, which show residence time from 150 years to 700 years.

Residence Time (Years)	Author	Year
700	Allen	2009
610	Zickfeld	2013
500	Matthews	2008
300	Plattner	2008
270	Cao	2010
230	Zickfeld	2012
220	Solomon	2012
220	Knutti	2012
210	Gillett	2011
180	Frolicher	2010
150	Hare	2006

https://agupubs.onlinelibrary.wiley.com/doi/abs/10.1002/2017JD028121

Assumptions

Keep current carbon emissions rise at 0.3 gt/yr (current)

Reduction in 45% of fossil fuel emissions by 2030 Decreases of carbon emissions will be offset by increases in population Atmospheric CO_2 stays the same slope. (Not increasing). However, rate of rise is increasing. Current rate is almost 3 ppm increase per year. At 100 years no more oil so CO_2 emissions drop by 55% Atmospheric CO_2 lowers to a minimum at year 2650 and then increases. We never reach equilibrium.

Even at a residence time of 100 years, atmospheric CO_2 never lowers.

Constraints for this graph. 45% reduction in fossil fuel CO_2 emissions by 2030 55% reduction in fossil fuel CO_2 emissions by 2130 due to depletion of those fuels. 2030

45% reduction in the rate of rise of Atmospheric CO_2. 2130 45% reduction in CO_2 concentration 2230 55% reduction in CO_2 concentration and rate.

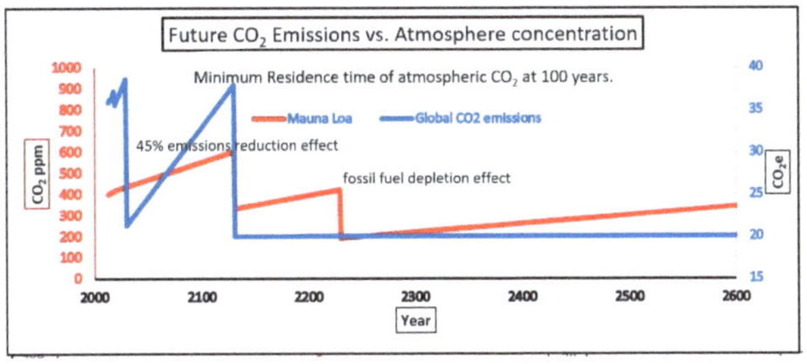

This is because we have massive loss of photosynthesis consumption. Globalforestwatch.org/map

Another way to look at residence time is a signature from past events, which lowered CO_2 emissions. For example, the oil embargo in the 1970's, multiple recessions and the big worldwide recession in 2009. The current COVID-19 pandemic. These are examples of lowered worldwide emissions. Below is the current graph of Mauna Loa CO_2. You can clearly see no signature from these events.

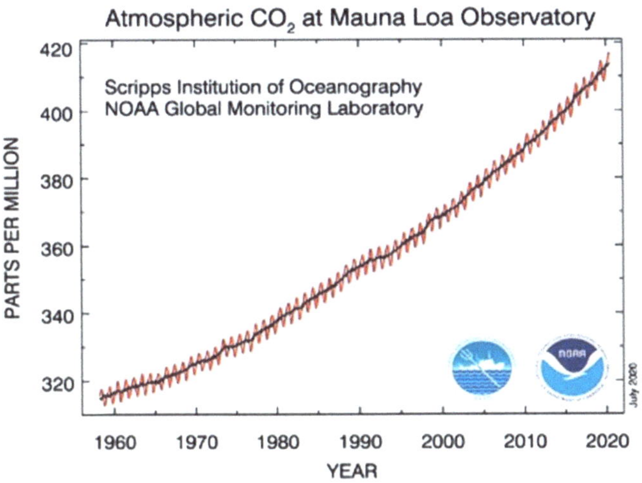

On Netflix, please watch "kiss the ground" movie. It clearly explains why we cannot lower atmospheric CO_2 by working on emissions of CO_2.

Sea Level Rise (or lack thereof)

https://www.ipcc.ch/report/sixth-assessment-report-working-group-i/ Twenty Ph. D's and I uploaded comments on Working Group 1 second order draft for AR6.
https://cctruth.org/comments_ar6wg1_sod.xlsx was uploaded 4/30/2020.
Sea Level Change data is unreliable. The satellite NOAA uses, (the Jason-3) has a minimum resolution of 25 mm. They say they are measuring a 3mm rise per year by measuring a location every 10 days. When we measure anything below minimum resolution, the data reliability drops exponentially below 50% of the minimum resolution. I put them in the document review for WG I AR6 for next year. I know the tide gauges tell the truth and show almost no sea level change. DOI : doi.org/10.33140/JMSRO.02.01.06
Review Article The Views of Three Sea Level Specialists, Mörner NA,
Wysmuller T and Parker
A https://www.opastonline.com/jmsro-volume-2-issue1-year-2019/www.opastonline.com J Mari Scie
Res Ocean, 2019 Volume 2 | Issue 1 See this document:

A movie called **Climate Hustle II** will come out October 2020 and show this.

In addition, the European satellite has a 1 mm minimum resolution and it shows the same sea level rise as the tide gauges at 1.06 mm/yr

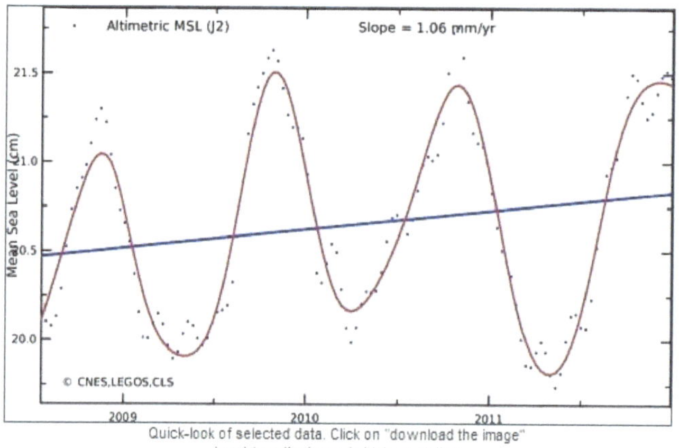

Quick-look of selected data. Click on "download the image" to retrieve the image in high resolution or on "download the data" to get the data.

The Jakobshavn Glacier in Greenland has grown for the third year in a row. This is the large one Al Gore and others have falsely said would melt and cause the oceans to rise 15 feet. https://earthobservatory.nasa.gov/images/145185/maj or-greenland-glacier-is-growing Tide gauge data:

https://sealevel.info/MSL_weighted.php?g_date=1910/ 1-2019/12&c_date=1910/12019/12&s_date=1910/12019/12&id=154,%20202,%20155,%20163,%20158,%20 188,%2012

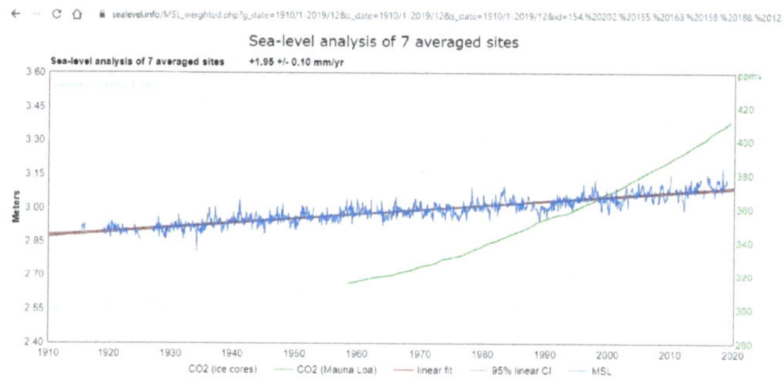

The mean sea level (MSL) trend is +1.95 mm/year with a 95% confidence interval of ±0.10 mm/year, based on monthly mean sea level data from 1915/5 to 2018/12. That is equivalent to a change of 0.64 feet in 100 years. (R-squared = 0.805)

Weight	ID	PSMSL	Station name
0.1429	270-061	154	Trieste, Italy
0.1429	170-161	202	Newlyn, UK
0.1429	1612340	155	Honolulu, HI, USA
0.1429	840-011	163	Balboa, Panama
0.1429	9410170	158	San Diego, CA, USA
0.1429	8724580	188	Key West, FL, USA
0.1429	8518750	12	The Battery, NY, USA

The plot shows the monthly mean sea level without the regular seasonal fluctuations due to coastal ocean temperatures, salinities, winds, atmospheric pressures, and ocean currents. By default, the long-term linear trend is also shown, in red, along with its 95% confidence interval. The plotted values are relative to the most recent Mean Sea Level datum established by NOAA CO-OPS or PSMSL.

Ocean Acidity

Ocean acidity (or lack thereof. Tony Heller shows how the ocean acidity is the same as it's always been in this video. Ocean stupidifcation

Net Zero

The document uses a term *Net Zero* with no definition.

We wrote the world's first atmospheric CO_2 equilibrium manuscript is peer reviewed and published in worlds top climate change journal by impact factor. Equilibrium Paper NetzeroCO2e=8.6gt/yr.

Truth about Al Gore

Web search "Club of Rome". This will tell you everything you need to know about the ignorance of Al Gore.

The assertion that 97% of scientists agree with the IPCC is wrong! This high consensus was touted because the three hundred manuscripts published between 2009 and 2013 were chosen for review on the basis of their seeming conformity to a certain point of view. Rejected for the review and survey of scientists were the more than seven hundred manuscripts written by scientists who had different statistics and conclusions from the ones that were wanted. Therefore, the agreeing part is 33%. We are 67%ers.

Discovery: Reduction in

Photosynthesis Correlation to Atmospheric CO_2 Increase. 65 more conferences have invited me to present this. I have not accepted any invites because we have no funding.

I sent these statistics to all 220 IPCC scientists by email.

Not one of them objected to the statistics. Atmospheric CO_2 is a binary system statistically. The two causes are CO_2 emissions and loss of photosynthesis. Each cause is multi-variate. We have had mostly flat human emissions (0.3 GT/yr vs. 0.6 GT/yr) since 2014. However, atmospheric CO_2 is still going up, and the rate of rise is increasing. In 2018, the Rxy correlation coefficient was 0.73 and not statistically significant (not cause and effect). In 2019 it is now 0.63 and dropping. The data is here:

Carbon Dioxide Does Not Freeze in the Atmosphere In the mesosphere, the pressure is 1 millibar. At this pressure, CO_2 freezes at -100°C. The temperature in the mesosphere is -90°C.

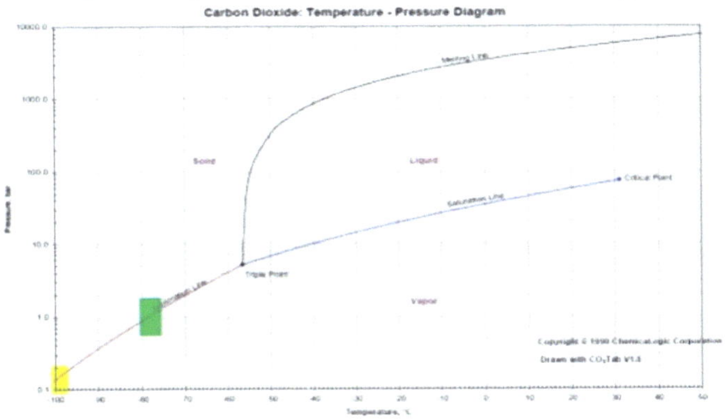

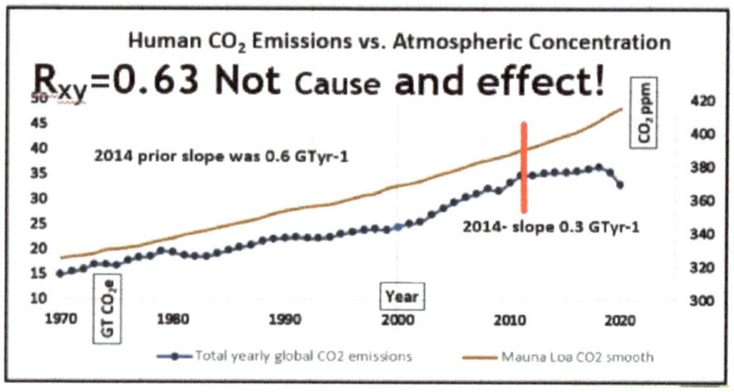

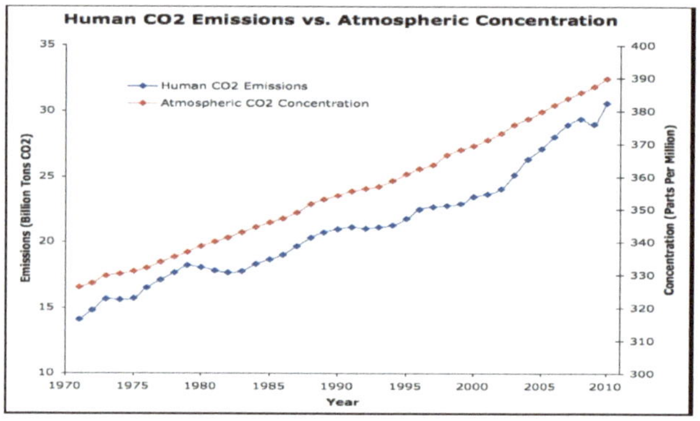

This 2010 graph is the only one you will see online. They do not want you to know how emissions of CO_2 have slowed down worldwide.

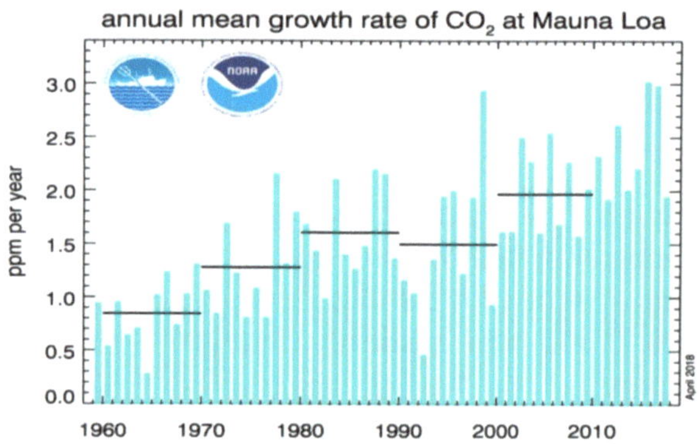

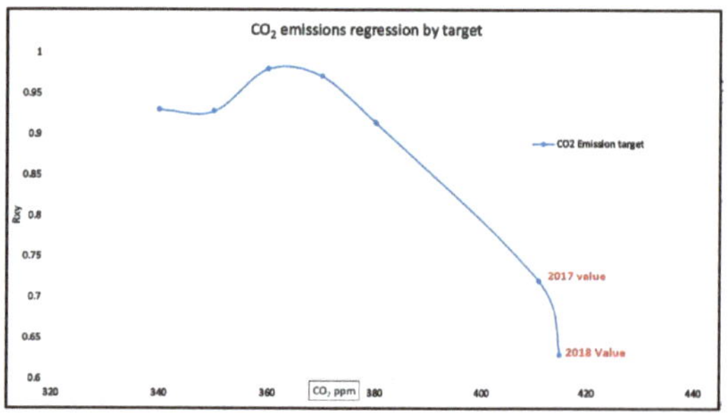

Carbon dioxide emissions correlate to 363 ppm and is a contributor, not the cause of the rise.

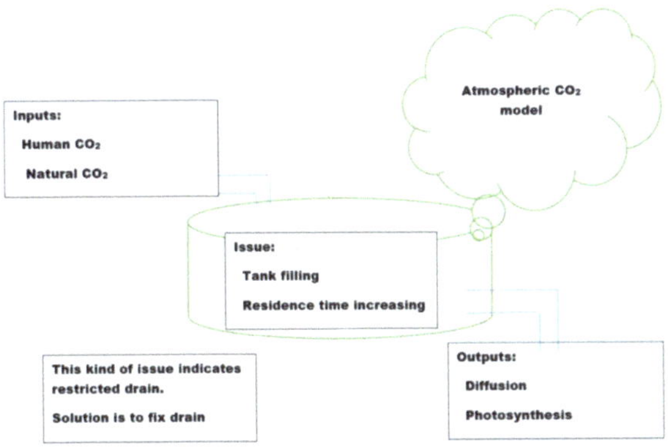

This tank model is like your kitchen sink. Standing water in the sink is increasing residence time. By this model, we need to shut the input and fix the drain. We cannot shut the input because the "natural" emissions are 20 billion tons/yr. We must increase photosynthesis.

The oscillation at Mauna Loa starts as a very strong signal in South America and then fans out larger and larger until Barrow's Alaska. The countries in South America burn the Amazon Rainforest, the densest forest in the world, from October/ November through May of the next year. Since 1950, an average of 30 million acres per year have been deforested and burned. So much CO_2 has been released that the trees and plants have grown too fast and died. This massive decay is what caused the Amazon Rainforest to switch to an oxygen sink and carbon dioxide producer. Hundreds of papers have been published on this.
Currently, the Amazon output is 15 $GTyr^{-1}$ of CO_2.

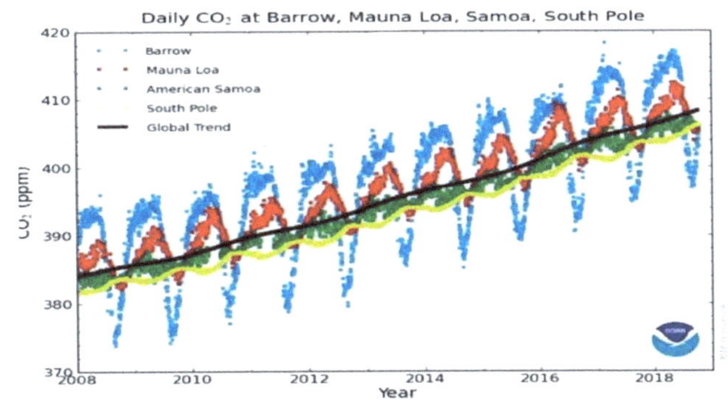

Mauna Loa cycles

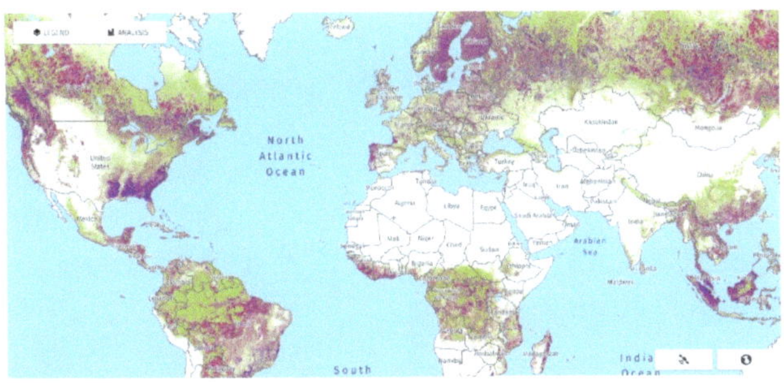

globalforestwatch.org 390->8.6 gtyr-1

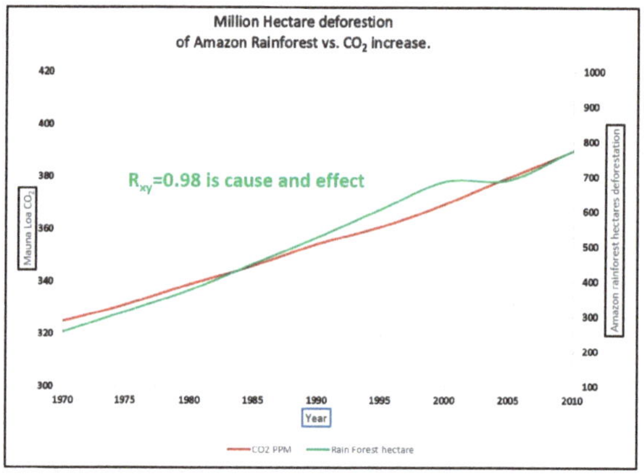

The Amazon Rainforest deforestation is a 0.98 cause and effect to the rise of carbon dioxide since 1957.

Amazon Rainforest R_{xy} =-0.99 The loss of oxygen worldwide is a 0.99 cause and effect to the destruction of 2 billion acres of the Amazon Rainforest since 1950! The correct solution is to stop non-sustainable deforestation of those forests like the Indian and Amazon Rainforests and plant 200 billion native trees and shrubs.

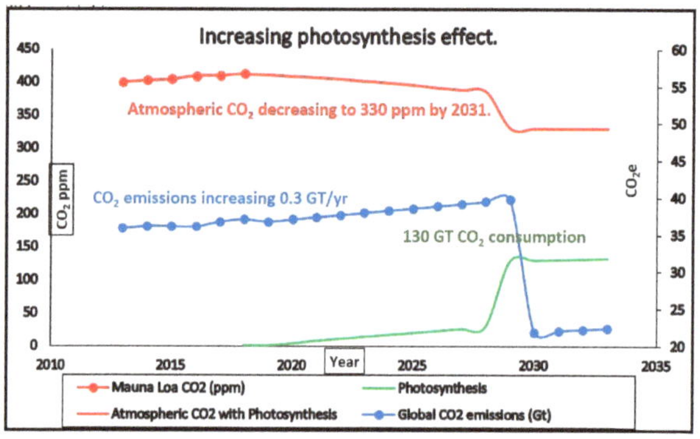

India stopped deforestation and is planting trees!
China is planting billions of trees!
Pakistan planted 1 billion trees in 2018, 2 billion more in 2019, and they will plant 8 billion more in the next four years! Peru stopped deforestation in 2020! Already planting 3 billion trees and the global garden greening atmospheric CO_2 minimum on October 4th was 407.51 ppm. Dr Pieter Tans said it should be 408.6+/- 0.5. For November the rise was -0.45 ppm. (11/1= 411.02, 4/20=410.57), November of 2017 it was 2.7 ppm rise. November 2018 1.85 ppm rise. 8 billion more trees scheduled in the next 4 years. We can easily plant 100 billion trees in the USA and in 10 years will consume an extra 10 billion tons annually.

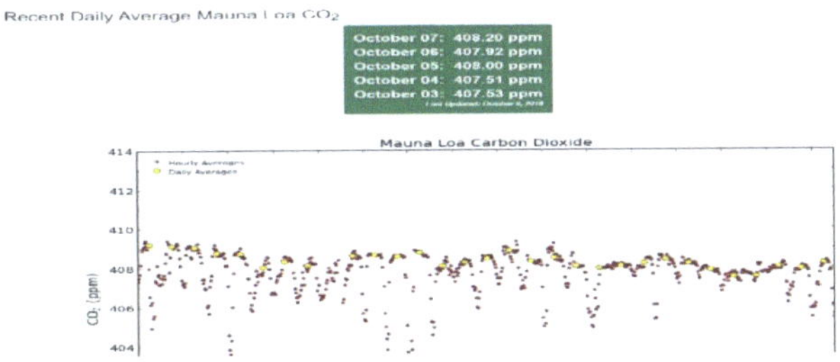

Effect of 24+ billion trees planted in the last 48 months.

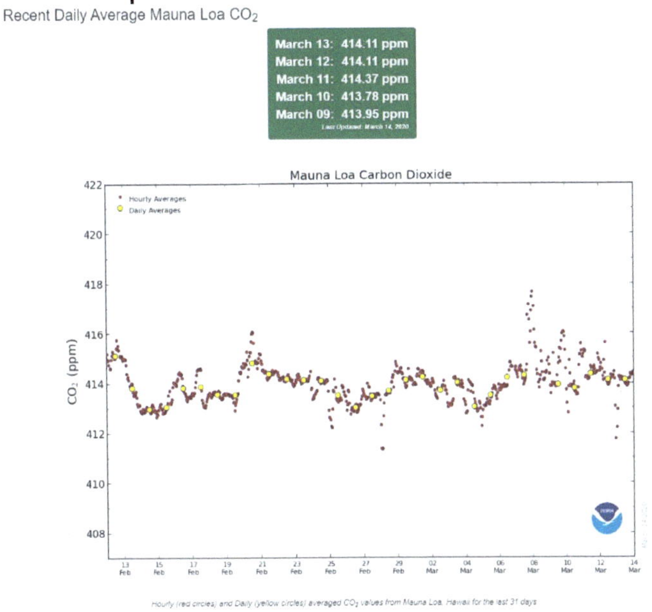

This drone can plant 40,000 trees per day.

I put in a complaint to Department of Commence Inspector general about Mauna Loa CO_2 fraud. They started investigating 4/24/20. Please download the rain-forest stop document and follow it weekly. Over 1000 people have been doing this since last June. To lower atmospheric Carbon dioxide quickly.

1. **Put pressure on Brazil and other Amazon rain-forest countries to stop deforestation ASAP. Also stop the biomass burning that puts 300 million tons of carbon dioxide into the atmosphere each year. This has caused 50ppm of the recent rise in**

atmospheric carbon dioxide concentration. Then after 10 years finish burning what is needed at 10% per year for 10 years.

2. Provide space where public can come and plant trees and shrubs. All government-owned lands. Very small cost. Need website with document for each planting area.
3. Plant shrubs in all freeway medians and sides. This is revenue plus in a two-year cycle. Plant native shrubs at a minimal spacing so all light is used in photosynthesis. This will take in 1 ton of CO_2 emissions per acre per year right at the source. The space would not need to be mowed every week in the summer.
4. Get schools involved and planting massive number of trees and shrubs. In their property and the government property as in 1 above.
5. Parks can add trees and shrubs.
6. Close any climate change research group. Not needed, unless doing photosynthesis work.
7. Tax incentive for business to plant trees and shrubs.
8. Wild fire attention. Get a retainer for the 747 plane and use it from the start on any wild fire.

Forest management by "strip logging" which was developed by Oregon State Forestry. This strip 30 to 60 yards wide (depending on the height of the trees) will provide ongoing logging opportunities, making these cuts. The side trees and shrubs will naturally reseed these cuts. These seeds are matched genetically to the local soil and climate. They grow much faster because of this. No reseeding is needed or desired. These cuts make an excellent firebreak.

We have an experiment on US 26 eastbound just west of Portland, Oregon. A permit obtained from Oregon Department of Transportation. These sensors are NIST certified and calibrated within one part per million. Graph 9 shows the rate of rise of atmospheric carbon dioxide less than 3 ppm/yr. The blue line represents the difference between the treed area and a non-treed area. Each location has a wind directional measurement. This measurement can confirm bad data from crosswind for example. This experiment proves we can plant native shrubs and trees by roads and freeways instead of grass. This freeway has 161,000 autos per day on it, and approximately 460 auto exit (Sylvan exit 71) per day between the two sensor locations. The final day of testing was 6/12/2021.

Procedure:

Place sensors at 6am daily for two weeks every other month for one year.
Pick up sensors at 7pm and analyze the data.
Put SD memory card from sensor into computer. Import the data into an Excel spreadsheet.

Repeat for other sensor.

For each 10 seconds subtract the treed area from the non-tree area.

Sort data for "smallest to largest" from subtraction result.

Remove negative numbers in the subtraction result.

The negative numbers are from wind gusts. We tracked this many times.

Calculate average for the day.

Things to note in the graph. At no time did the blue line go below the red line. On December 20th, a very dark and rainy day the difference was 9 ppm. In April through June we had very little rain. The graph shows this as lower difference. For photosynthesis, we need these things, light, vegetation, moisture and carbon dioxide. Experiment Summary: This experiment proves we can plant native trees and shrubs instead of grass and they will eventually consume all the carbon dioxide from the vehicles. This is applicable for ±50° from the equator.

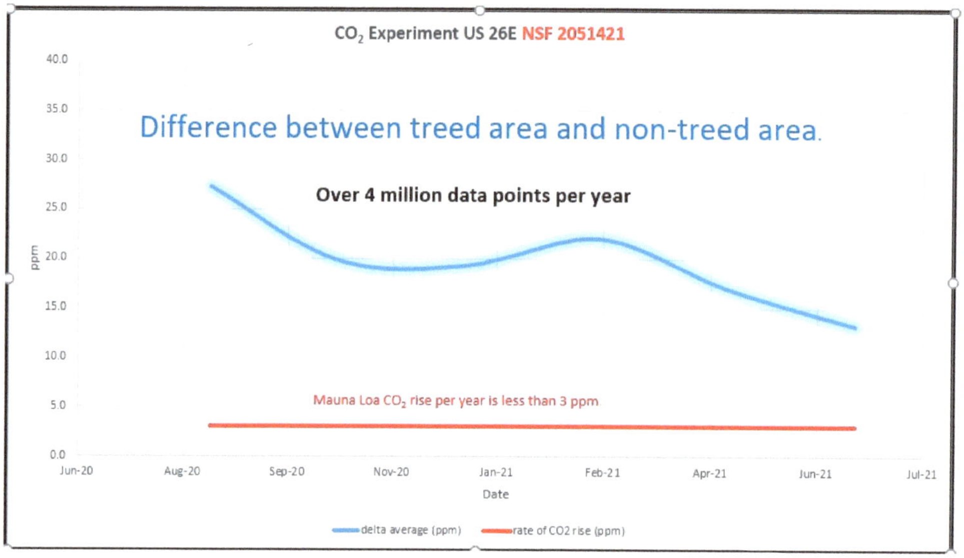

$$6CO_2 + 6H_2O + \lambda \rightarrow C_6H_{12}O_6 + 6O_2$$

I contacted the National Academy of Sciences, spoke to Dr. Mike Kuperberg who is the Executive Director of the U.S. Global Change Research Program (USGCRP), He saw the correct science in our presentations page. He sent it to the other scientists in their office. Their consensus was to have me get a team and participate in the annual "Expert and Government Review (EAGR)" program of the Intergovernmental Panel on Climate Change (IPCC) reports.

I led a team of PhD's whose ranks soon swelled from myself to thirty other scientists who are also participating in the writing of this college textbook. Together we participated in the "EAGR" program, and we unanimously found all kinds of garbage science in their reports. Also, we had Adam Yeeley, the chief editor of Nature Climate Change fired. His PhD was in political science. He let the IPCC scientists publish loosely referenced manuscripts and circular reference them in their reports. This is not science.

The IPCC reports are deliberate science fiction. The IPCC writers identify themselves as climate experts and inform governments globally in their reports on what to believe about climate change. These false reports lead to false government policies being made that negatively impact every person and business around the globe through unnecessary economic restrictions and taxation.

In our PhD review of IPCC working Group 1, in the first order draft for Ar6 we found their inaccurate global warming potential model. This model assumes equal greenhouse gas (GHG) concentrations. This equal concentration will never happen in reality. Carbon dioxide is more than 200 times the concentration of methane. Furthermore, we found in Annex 2, a table with the correct order of GHG effects. Any model which ignores data to benchmark it with is an inaccurate model. We sent our review at least 23 times to them to correct their inaccuracies and they ignored our scientific finding. That makes the AR6, report worthless as a whole. However, for the final draft for AR6 they deleted the table from Annex 2! Instead of making changes to make their model they deleted the benchmarking data in Annex 2. This is how corrupt they are. You can't have an accurate model without benchmark data to validate it.

Disclaimer: Sometimes the IPCC changes things without notification. For example, the Executive Summary of the Mitigation Chapter had our review paragraph added. However now to confuse people they start out every paragraph the same. Previously this was not done. Also they changed the numbering scheme for the chapters. The difference is they are now beginning four paragraphs with this statement, "Limiting warming to 1.5°C depends on greenhouse gas (GHG) emissions". The three paragraphs that start with this statement have nothing to do with our review and are just there to mislead people. In fact, they still state inaccuracies they've been told about on several occasions such as methane gas is the worst greenhouse gas. However, by scientific measurement, it is clear that methane gas is 0.29% effect and water vapor is 89.4% greenhouse gas effect. See Chapter 2.

In our 23-30 scientific PhD review of IPCC working Group 1 first order draft for Ar6 we found their faulty global warming potential model. This model assumes equal greenhouse gas (GHG) concentrations. This equal concentration will never happen in reality. For example, carbon dioxide is more than 200 times the concentration of methane. Furthermore, in Group 1, we found in Annex 2, a table with the correct order of GHG effects. Any model which ignores data to benchmark it with this correct order is a fake model. We sent our review at least 23 times to inform them they had to benchmark their Annex 2 table to the correct order of GHG effects. However, for the final draft for Ar6 they chose not to benchmark their final draft but instead chose to delete the table in Annex 2, which still left their fake GWP model intact. This wasn't just overlooking the benchmarking of the data. They purposely hid the fact that their science model was false. This is how corrupt they are.

Twenty-three to thirty PhD's participate in "Expert and Government review" program for the IPCC reports. We find all kinds of garbage in them. Each member of our team downloads the reports by various "working groups" such as the IPCC. We go through those reports line by line. Then we have an online meeting and decide what we will submit for changes. Then we each submit the same changes twenty-three to thirty times.

For example, for their mitigation chapter, Jim Skea said we need to lower atmospheric carbon dioxide emissions by 45% by 2030. However, the statement in the chapter he was basing that goal on was buried on page 95 and had no references (citations). They completely made it up! Also buried on page 101 was a statement stating that the probability of their solution to work is 66%. When we submit our review, they put these things in the 5th paragraph of their executive summary.

https://www.ipcc.ch/report/ar6/wg3/downloads/report/IPCC_AR6_WGIII_FullReport.pdf

Our team of 30 scientific PhD's forced working group III to move the statement with no references (citation) from page 95 to page 6 paragraph B.1.3

B.1.3 Historical cumulative net CO2 emissions from 1850 to 2019 were 2400 } 240 GtCO2 (high confidence). Of these, more than half

(58%) occurred between 1850 and 1989 [1400 } 195 GtCO2], and about 42% between 1990 and 2019 [1000 } 90 GtCO2]. About

17% of historical cumulative net CO2 emissions since 1850 occurred between 2010 and 2019 [410 } 30 GtCO2].10 By comparison,

the current central estimate of the remaining carbon budget from 2020 onwards for limiting warming to 1.5°C with a probability of 50% has been assessed as 500 GtCO2, and as 1150 GtCO2 for a probability of 67% for limiting warming to 2°C. Remaining carbon budgets depend on the amount of non-CO2 mitigation (}220 GtCO2) and are further subject to geophysical uncertainties. Based on central estimates only, cumulative net CO2 emissions between 2010 and 2019 compare to about four-fifths of the

size of the remaining carbon budget from 2020 onwards for a 50% probability of limiting global warming to 1.5°C, and about one-third of the remaining carbon budget for a 67% probability to limit global warming to 2°C. Even when taking uncertainties into account, historical emissions between 1850 and 2019 constitute a large share of total carbon budgets for these global warming levels.11,12 Based on central estimates only, historical cumulative net CO2 emissions between 1850 and 2019 amount to about four-fifths12 of the total carbon budget for a 50% probability of limiting global warming to 1.5°C (central estimate about 2900 GtCO2), and to about two thirds12 of the total carbon budget for a 67% probability to limit global warming to 2°C (central estimate about 3550 GtCO2). {Figure 2.7, 2.2, Figure TS.3, WGI Table SPM.2}

Exhibit II

IPCC

The Intergovernmental Panel on Climate Change Ignores Key Data, Simulation Results are invalid cctruth.org

SUMMARY

The Intergovernmental Panel on Climate Change reports are inaccurate and are falsely skewing Data. Publishing garbage manuscripts in a journal whose chief editor has a PhD in Political Science. There reports are deliberate scientific fiction. https://cctruth.org/ipcc.pdf This is well documented with links to their reports and descriptions where we found the items.

IPCC Reports

The IPCC cherry-picks the relatively few reports which follow and support their own agenda, rejecting the greater number of reports that do not support that agenda. They have ignored the oppositional findings of more than one thousand reports about the Amazon Rainforest. Any scientist who cherry-picks data would be shamed out of a job. More than 60% of the references in their reports were to the previously farce Journal Nature Climate Change who had as Chief Editor Adam Yeeley. His Ph.D is in Political Science. He let scientists publish garbage manuscripts so they could circular reference them in the IPCC reports. This is not science! He is just there to keep correct science out and publish crap science. However, after sending email, to their board he is no longer there. Still that journals manuscripts reference the IPCC reports. The IPCC reports then reference the manuscripts in that journal. Circular referencing is not science! June 2020 I notified the board of this and they fired him the next day. Bronwyn Wake is the board member who took Adam's place. Initially they said she was chief editor for many years prior to June of 2020. I complained and they changed when she started to June 2020. The kind of garbage getting published was like the manuscript in early July which said the Antarctic was warming. This was all over the worldwide news for a few days. This garbage manuscript like the reset under Adam had the title and abstract matched, however they didn't match the manuscript. The manuscript said the warming was a 20-year cycle that started in 2020 and is cooling now!

We performed an expert review of IPCC (Intergovernmental Panel on Climate Change) SR 1.5 Chapter Two "Mitigation" .https://cctruth.org/expert_review_SR1.5_mitigation.pd f . These are the key findings: Their equilibrium statements had no references to any published manuscripts. One of the chapter scientists replied and said they are not equilibrium statements and they are from simulations. I showed their simulations to a friend who has 27 years' experience and he started uncontrollable laughter. Further down in their document was the only probability they did is 50-66% for their solution by lowering emissions will work. I sent this to around 1000 scientists, the worldwide media, the UN and IPCC scientists. The media ignored it, however, IPCC working Group 1 and 3 saw my expert review ability and invited us to review their reports for AR6 next year. https://cctruth.org/comments_ar6wg3_fod.xlsx is already accepted for WG 3.

https://cctruth.org/comments_ar6wg1_sod.xlsx was uploaded 4/30/2020.

2019 IPCC SR 1.5 Chapter 2 "Limiting warming to 1.5°C depends on greenhouse gas (GHG) emissions over the next decades, where lower GHG emissions in 2030 lead to a higher chance of keeping peak warming to 1.5°C (high confidence). Available pathways that aim for no or limited (less than 0.1°C) overshoot of 1.5°C keep GHG emissions in 2030 to 25–30 GtCO2e yr−1 in 2030 (interquartile range). This contrasts with median estimates for current unconditional NDCs of 52–58 GtCO2e yr−1 in 2030

(https://www.ipcc.ch/sr15/chapter/chapter-2/, Page ES, 5th paragraph). Now their Executive Summary

(https://cctruth.org/es.pdf) shows this statement with no references and their probability of 66%. I sent four emails asking them where these numbers came from. A research scholar at The International Institute for Applied Systems Analysis (IIASA) Schlossplatz 1, A-2361 Laxenburg, Austria replied: "Dear Dave, Thank you very much for your question on the assessment of quantitative pathways in the SR15. The statement is taken from Table 2.4, bottom section, third row, first column, rounded to multiples of 5. The assessment in this table is based on the ensemble of quantitative pathways compiled by the IAMC and IIASA for the IPCC SR15 process

(https://doi.org/10.22022/SR15/08-2018.15429). The Python script for preparing this table is available under an open-source license at https://data.ene.iiasa.ac.at/sr15_scenario_analysis/asse

ssment/sr15_2.3.3_global_emissions_statistics.html (see https://doi.org/10.22022/SR15/08-2018.15428 for the scientific reference of the assessment notebooks).

Neither the statement nor the table does make any assertion about an equilibrium; it is merely an assessment of the pathways at a specific point in time [bold added]. I do hope that this clarifies your request. The International Institute for Applied Systems Analysis (IIASA) Schlossplatz 1, A-2361 Laxenburg, Austria." Please note! This faulty simulation has us reach equilibrium at 2050!

I looked at their simulations and they are garbage because they don't have boundary conditions. Their simulation shows NetZero at zero to in 2050. However, the IPCC and UN have started this false 12 year doomsday garbage. This is why nothing they have predicted has or will come true. Dr. Kevin Dayaratna testified at the Oregon Carbon group with the correct use of their simulations. https://ctruth.org/DAYARATNA.mp4

Earlier I sent this review to 5000 scientists and all the worldwide media by email with delivery and read receipts. They read it. One NOAA scientist replied and said I should go after the publishers of the IPCC crappy manuscripts. I thanked him and said I would if I had a large staff of scientists. I showed their simulations to an expert in simulations and he started uncontrollable laughter. Around December 15th 2019 I sent it to all other than Chapter two IPCC scientists. Our review was sent to the other 200 IPCC scientists who essentially agreed with the review we provided.

Rare Use of Probability

"For limiting global warming to below 2°C with at least 66% probability [bold added] CO2 emissions are projected to decline by about 25% by 2030 in most pathways (10–30% interquartile range) and reach net zero around 2070 (2065–2080 interquartile range).1 {2.2, 2.3.3, 2.3.5, 2.5.3, Cross-Chapter Boxes 6 in Chapter 3 and 9 in Chapter 4, 4.3.7} (p 21.3, Table 2.1).

"No pathways were available that achieve a greater than 50-66% probability of limiting warming below 1.5° C [bold added] during the entire 21st century based on the MAGICC model projections" For limiting global warming to below 2°C with at least 66% probability CO2 emissions are projected to decline by about 25% by 2030 in most pathways (10–30% interquartile range) and reach net zero around 2070 (see p. ES, Paragraph 5). The probability is actually zero because the minimum residence time is hundreds of years. (Probability Table 2.1 page 21.3)

(No business would spend such a significant amount of money (2.8 trillion dollars already spent worldwide) on a project with only a 50-66% chance of success.) Their probability is actually zero because the average residence time for atmospheric CO2 is 150 years. (IPCC 2003)

Citation

"This chapter should be cited as: Rogelj, J., D. Shindell,

K. Jiang, S. Fifita, P. Forster, V. Ginzburg, C. Handa, H. Kheshgi, S. Kobayashi, E. Kriegler, L. Mundaca, R.

Séférian, and M.V.Vilariño, 2018: Mitigation Pathways Compatible with 1.5°C in the Context of Sustainable

Development. In: Global Warming of 1.5°C. An IPCC Special Report on the impacts of global warming of 1.5°C above pre-industrial levels and related global greenhouse gas emission pathways, in the context of strengthening the global response to the threat of climate change, sustainable development, and efforts to eradicate poverty [Masson-Delmotte, V., P. Zhai, H.-O. Pörtner, D. Roberts, J. Skea, P.R. Shukla, A. Pirani, W.

Moufouma-Okia, C. Péan, R. Pidcock, S. Connors, J.B.R. Matthews, Y. Chen, X. Zhou, M.I. Gomis, E. Lonnoy, T. Maycock, M. Tignor, and T. Waterfield (eds.)]. In Press" (p. 93)

Use of Unscientific Terms

The document uses the unscientific terms highly (or otherwise) likely six times, unlikely three times, and highly (or otherwise) confident sixty-two times. In every case, percent probability must be used.

Planting Native trees is the only way to lower Atmospheric carbon dioxide to 330 ppm by 2031.

The IPCC follows a false agenda and a false GWP (Global Warming Potential) Calculation, neither of which is based on reality. Their GWP calculation assumes equal greenhouse gas concentrations of methane, nitrous oxide and carbon dioxide and other gases, which will never happen in reality. If we did have equal concentrations of N2O (laughing gas) for instance, the people in the world would have silly smiles on their faces and high-pitched voices. IPCC Working group I, second order draft (SOD) Annex II found 14 published manuscripts which show the same data as Dr. Blasings. These were published prior to the GWP and the IPCC ignored them. We put this finding in our review for working group 1. They ignored it and deleted the 14 manuscripts! Any model which is not verified by data is a false model. The correct order of greenhouse gases CO2 then CH4 then N2O then NO (highest effect to lowest effect) Dr. TJ Blasing exposed the greenhouse gases with longwave radiation and was thus able to calculate the actual effect.

http://cctruth.org/index.php/ghg/ Methane is 0.5 watts/m2. CO2 is 1.94 watts/m2. The media should not believe the IPCC or the UN when it comes to climate change. Dr. Hal Dorian passed away 4/28/20. His memorial. He is one of the NASA scientists who helped write our proposal. We dedicate our proposal to him.

Planting trees is 100% probability to lower atmospheric carbon dioxide.

Residence Time of Atmospheric CO2

Residence time is how long a molecule will stay in a location before being released. Like standing water in your kitchen, sink. The water is residing longer. A 2003 IPCC report shows residence time increased from 5 to 200 years. Dr. TJ Blasing shows 100-300 years. In 2016, I emailed Dr. Jim Hansen and two other prominent climate-change scientists that emissions had been flat since 2014, but that atmospheric CO2 was still increasing and the rate of rise was still increasing. I asked them how this could be happening--if emissions were the cause of atmospheric CO2 increase. They said we must wait another 470 years for anything we do with emissions to show an effect. Anything we do with CO2 emissions has not and will not have any effect on atmospheric CO2 for hundreds of years. However, the residence time for atmospheric carbon dioxide is 150 years. This is why everything we have done to lower emissions of CO2 has had zero effect on the atmospheric CO2 rise.

https://cctruth.org/residence_time.pdf Below are the constraints I used. Even at average residence time of 100

years Mauna Loa never stays low.

Facts

Residence time was 5 years, Now more than 150 years. Recently I sent out a survey email to 400 climate change scientists about atmospheric CO2 residence time. Most scientists said 200-400 years. One scientist sent me his research of published papers, which show residence time from 150 years to 700 years.

Residence Time (Years)	Author	Year
700	Allen	2009
610	Zickfeld	2013
500	Matthews	2008
300	Plattner	2008
270	Cao	2010
230	Zickfeld	2012
220	Solomon	2012
220	Knutti	2012
210	Gillett	2011
180	Frolicher	2010
150	Hare	2006

https://agupubs.onlinelibrary.wiley.com/doi/abs/10.10 02/2017JD028121

Assumptions

Keep current carbon emissions rise at 0.3 gt/yr (current)

Reduction in 45% of fossil fuel emissions by 2030 Decreases of carbon emissions will be offset by increases in population Atmospheric CO2 stays the same slope. (Not increasing). However, rate of rise is increasing. Current rate is almost 3 ppm increase per year. At 100 years no more oil so CO2 emissions drop by 55% Atmospheric CO2 lowers to a minimum at year 2650 and then increases. We never reach equilibrium.

Even at a residence time of 100 years, atmospheric CO2 never lowers.

Constraints for this graph. 45% reduction in fossil fuel CO2 emissions by 2030 55% reduction in fossil fuel CO2 emissions by 2130 due to depletion of those fuels. 2030 45% reduction in the rate of rise of Atmospheric CO2. 2130 45% reduction in CO2 concentration 2230 55% reduction in CO2 concentration and rate.

This is because we have massive loss of photosynthesis consumption. Globalforestwatch.org/map

Another way to look at residence time is a signature from past events, which lowered CO2 emissions. For example, the oil embargo in the 1970's, multiple recessions and the big worldwide recession in 2009. The current COVID-19 pandemic. These are examples of lowered worldwide emissions. Below is the current graph of Mauna Loa CO2. You can clearly see no signature from these events.

On Netflix, please watch "kiss the ground" movie. It clearly explains why we cannot lower atmospheric CO2 by working on emissions of CO2.

Sea Level Rise (or lack thereof)

https://www.ipcc.ch/report/sixth-assessment-report-working-group-i/ Twenty Ph. D's and I uploaded comments on Working Group 1 second order draft for AR6. https://cctruth.org/comments_ar6wg1_sod.xlsx was uploaded 4/30/2020.

Sea Level Change data is unreliable. The satellite NOAA uses, (the Jason-3) has a minimum resolution of 25 mm. They say they are measuring a 3mm rise per year by measuring a location every 10 days. When we measure anything below minimum resolution, the data reliability drops exponentially below 50% of the minimum resolution. I put them in the document review for WG I AR6 for next year. I know the tide gauges tell the truth and show almost no sea level change. DOI : doi.org/10.33140/JMSRO.02.01.06 Review Article The Views of Three Sea Level Specialists, Mörner NA,

Wysmuller T and Parker

A https://www.opastonline.com/jmsro-volume-2-issue1-year-2019/www.opastonline.com J Mari Scie Res Ocean, 2019 Volume 2 | Issue 1 See this document:

A movie called Climate Hustle II will come out October 2020 and show this.

In addition, the European satellite has a 1 mm minimum resolution and it shows the same sea level rise as the tide gauges at 1.06 mm/yr

The Jakobshavn Glacier in Greenland has grown for the third year in a row. This is the large one Al Gore and others have falsely said would melt and cause the oceans to rise 15 feet. https://earthobservatory.nasa.gov/images/145185/maj or-greenland-glacier-is-growing Tide gauge data:

https://sealevel.info/MSL_weighted.php?g_date=1910/ 1-2019/12&c_date=1910/12019/12&s_date=1910/12019/12&id=154,%20202,%20155,%20163,%20158,%20 188,%2012

Ocean Acidity

Ocean acidity (or lack thereof. Tony Heller shows how the ocean acidity is the same as it's always been in this video. Ocean stupidifcation

Net Zero

The document uses a term Net Zero with no definition.

We wrote the world's first atmospheric CO2 equilibrium manuscript is peer reviewed and published in worlds top climate change journal by impact factor. Equilibrium Paper NetzeroCO2e=8.6gt/yr.

Truth about Al Gore

Web search "Club of Rome". This will tell you everything you need to know about the ignorance of Al Gore.

The assertion that 97% of scientists agree with the IPCC is wrong! This high consensus was touted because the three hundred manuscripts published between 2009 and 2013 were chosen for review on the basis of their seeming conformity to a certain point of view. Rejected for the review and survey of scientists were the more than seven hundred manuscripts written by scientists who had different statistics and conclusions from the ones that were wanted. Therefore, the agreeing part is 33%. We are 67%ers.

Discovery: Reduction in

Photosynthesis Correlation to Atmospheric CO_2 Increase. 65 more conferences have invited me to present this. I have not accepted any invites because we have no funding.

I sent these statistics to all 220 IPCC scientists by email.

Not one of them objected to the statistics. Atmospheric CO_2 is a binary system statistically. The two causes are CO_2 emissions and loss of photosynthesis. Each cause is multi-variate. We have had mostly flat human emissions (0.3 GT/yr vs. 0.6 GT/yr) since 2014. However, atmospheric CO_2 is still going up, and the rate of rise is increasing. In 2018, the Rxy correlation coefficient was 0.73 and not statistically significant (not cause and effect). In 2019 it is now 0.63 and dropping. The data is here:

Carbon Dioxide Does Not Freeze in the Atmosphere In the mesosphere, the pressure is 1 millibar. At this pressure, CO_2 freezes at -100°C. The temperature in the mesosphere is -90°C.

This 2010 graph is the only one you will see online. They do not want you to know how emissions of CO_2 have slowed down worldwide.

Carbon dioxide emissions correlate to 363 ppm and is a contributor, not the cause of the rise.

This tank model is like your kitchen sink. Standing water in the sink is increasing residence time. By this model, we need to shut the input and fix the drain. We cannot shut the input because the “natural” emissions are 20 billion tons/yr. We must increase photosynthesis.

The oscillation at Mauna Loa starts as a very strong signal in South America and then fans out larger and larger until Barrow's Alaska. The countries in South America burn the Amazon Rainforest, the densest forest in the world, from October/ November through May of the next year. Since 1950, an average of 30 million acres per year have been deforested and burned. So much CO_2 has been released that the trees and plants have grown too fast and died. This massive decay is what caused the Amazon Rainforest to switch to an oxygen sink and carbon dioxide producer.

Hundreds of papers have been published on this.

Currently, the Amazon output is 15 $GTyr^{-1}$ of CO_2.

Mauna Loa cycles

globalforestwatch.org 390->8.6 gtyr-1

The Amazon Rainforest deforestation is a 0.98 cause and effect to the rise of carbon dioxide since 1957.

Amazon Rainforest Rxy =-0.99 The loss of oxygen worldwide is a 0.99 cause and effect to the destruction of 2 billion acres of the Amazon Rainforest since 1950! The correct solution is to stop non-sustainable deforestation of those forests like the Indian and Amazon Rainforests and plant 200 billion native trees and shrubs.

India stopped deforestation and is planting trees!

China is planting billions of trees!

Pakistan planted 1 billion trees in 2018, 2 billion more in 2019, and they will plant 8 billion more in the next four years! Peru stopped deforestation in 2020! Already planting 3 billion trees and the global garden greening atmospheric CO2 minimum on October 4th was 407.51 ppm. Dr Pieter Tans said it should be 408.6+/- 0.5. For November the rise was -0.45 ppm. (11/1= 411.02, 4/20=410.57), November of 2017 it was 2.7 ppm rise. November 2018 1.85 ppm rise. 8 billion more trees scheduled in the next 4 years. We can easily plant 100 billion trees in the USA and in 10 years will consume an extra 10 billion tons annually.

Effect of 24+ billion trees planted in the last 48 months.

This drone can plant 40,000 trees per day.

I put in a complaint to Department of Commence Inspector general about Mauna Loa CO2 fraud. They started investigating 4/24/20. Please download the rain-forest stop document and follow it weekly. Over 1000 people have been doing this since last June. To lower atmospheric Carbon dioxide quickly.

1. Put pressure on Brazil and other Amazon rain-forest countries to stop deforestation ASAP. Also stop the biomass burning that puts 300 million tons of carbon dioxide into the atmosphere each year. This has caused 50ppm of the recent rise in atmospheric carbon dioxide concentration. Then after 10 years finish burning what is needed at 10% per year for 10 years.

2. Provide space where public can come and plant trees and shrubs. All government-owned lands. Very small cost. Need website with document for each planting area.

3. Plant shrubs in all freeway medians and sides. This is revenue plus in a two-year cycle. Plant native shrubs at a minimal spacing so all light is used in photosynthesis. This will take in 1 ton of CO_2 emissions per acre per year right at the source. The space would not need to be mowed every week in the summer.

4. Get schools involved and planting massive number of trees and shrubs. In their property and the government property as in 1 above.

5. Parks can add trees and shrubs.

6. Close any climate change research group. Not needed, unless doing photosynthesis work.

7. Tax incentive for business to plant trees and shrubs.

8. Wild fire attention. Get a retainer for the 747 plane and use it from the start on any wild fire.

Forest management by "strip logging" which was developed by Oregon State Forestry. This strip 30 to 60 yards wide (depending on the height of the trees) will provide ongoing logging opportunities, making these cuts. The side trees and shrubs will naturally reseed these cuts. These seeds are matched genetically to the local soil and climate. They grow much faster because of this. No reseeding is needed or desired. These cuts make an excellent firebreak.

We have an experiment on US 26 eastbound just west of Portland, Oregon. A permit obtained from Oregon Department of Transportation. These sensors are NIST certified and calibrated within one part per million. Graph 9 shows the rate of rise of atmospheric carbon dioxide less than 3 ppm/yr. The blue line represents the difference between the treed area and a non-treed area. Each location has a wind directional measurement. This measurement can confirm bad data from crosswind for example. This experiment proves we can plant native shrubs and trees by roads and freeways instead of grass. This freeway has 161,000 autos per day on it, and approximately 460 auto exit (Sylvan exit 71) per day between the two sensor locations. The final day of testing was 6/12/2021.

Procedure:

Place sensors at 6am daily for two weeks every other month for one year.

Pick up sensors at 7pm and analyze the data.

Put SD memory card from sensor into computer. Import the data into an Excel spreadsheet.

Repeat for other sensor.

For each 10 seconds subtract the treed area from the non-tree area.

Sort data for "smallest to largest" from subtraction result.

Remove negative numbers in the subtraction result.

The negative numbers are from wind gusts. We tracked this many times.

Calculate average for the day. Repeat.

Things to note in the graph. At no time did the blue line go below the red line. On December 20th, a very dark and rainy day the difference was 9 ppm. In April through June we had very little rain. The graph shows this as lower difference. For photosynthesis, we need these things, light, vegetation, moisture and carbon dioxide. Experiment Summary: This experiment proves we can plant native trees and shrubs instead of grass and they will eventually consume all the carbon dioxide from the vehicles. This is applicable for ±50° from the equator.

$6CO_2 + 6H_2O + \lambda \rightarrow C_6H_{12}O_6 + 6O_2$

The second year finished on 5/16/2022 with over 4 million more data points. This moved the experiment from Theory to Scientific Law!

Texas needs to find native shrubs to plant in these locations.

Native western Oregon plants.

Sweet bay

Photinia

Juniper

Knick

Leaf holly

Red twig Dogwood

Where to plant

Medians Photinia, Sweet bay, Leaf holly, Red twig Dogwood

On/Off ramps Photinia, Sweet bay, Juniper, Knick

References.

1. https://cctruth.org/comments_ar6wg1_sod.xlsx was uploaded 4/30/2020.
2. DOI:10.1007/s11069-012-0159-8
3. https://www.omicsonline.org/open-access/the-essential-role-of-photosynthesis-in-defining-net-zero-carbon-dioxide-emissions-for-equilibrium-calculations.pdf
4. https://cctruth.org/NOAA_Mauna.pdf
5. https://cctruth.org/ WGII_SOD_Ar6.xlsx
6. https://gml.noaa.gov/ccgg/trends/global.html
7.

Chapter 14

Important Videos

On Netflix, *Kiss the Ground* movie.

https://www.youtube.com/watch?v=lbDbA32fNek

On Netflix, *Seaspiracy.*

On YouTube, "Jo Nova - How to Destroy a Perfectly Good Electricity Grid in Three Easy Steps," https://www.youtube.com/watch?v=JYHX-Ib3Q5Q (Solar panels and windmills are not the solution.)

Tony Heller's "Leaders In The Race To Net Zero," https://www.youtube.com/watch?v=5RBYMH9CM5o (Pulling Back the Curtain on Junk Science: realclimatescience.com)

The Oceans are not a sink for Atmospheric Carbon Dioxide

Tony Heller's Ocean Stupidification https://www.youtube.com/watch?v=ylpDngGm5cQ

Predatory Journals are a Fabrication.

The American Meteorological Society (AMS), Plus One Journal and the Nature Climate Change (NCC) Journals started the predatory journals list website to distract from what the AMS and Plus One and the NCC are doing. These AMS, Plusone and NCC journals charge thousands of dollars to publish manuscripts based on jaundiced science, and then perform prejudiced peer reviews. Their chief editors operate on a “belief” system based on political narratives and not the climate science issues.

For example, if a manuscript doesn’t match their belief system, then it won't go to peer review. These journals also have a very high publishing rate. They almost always publish everything the chief editor allows to be peer reviewed. Also, their impact factors (the benchmark for journal effectiveness) are typically lower than the so-called predatory journals.

Predictably, the predatory journals list maligns many good journals, like the respected International Journal of Chemical Engineering. These journals generally charge lower rates in the range of $300 to $500 to publish a manuscript. Furthermore, they usually have higher impact factor scores than the AMS, Plus One or NCC journals. The Predatory Journals chief editors, like myself, function with an open-minded, scientific review system. That’s what actual scientists do and teach their students to do as well.

Chapter 15 Questions:

1. What is the impact factor for the NCC Journal?
2. What is the impact factor of some of the predatory journals on that list? List five journals and their respective impact factor.

Chapter 16 Antarctic Sea ice is Growing

The reason the Antarctic ice is growing is because of astrophysical warming you studied in chapter four. Year over year the Antarctic Sea Ice is growing (2). See Figure 2.

Figure 1

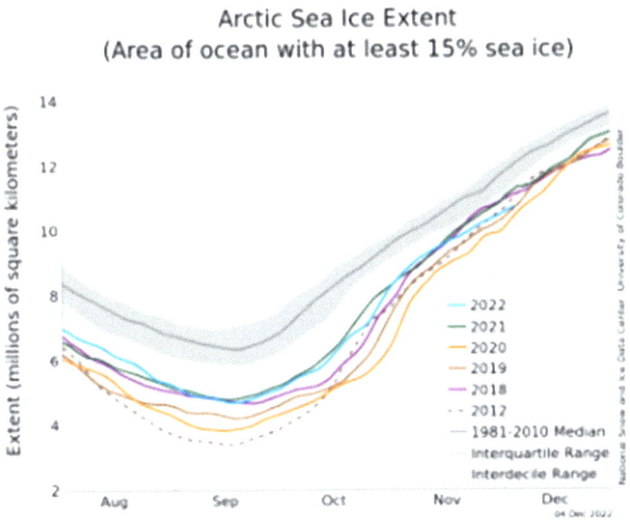

Figure 2.

References:

1. https://www.csmonitor.com/Science/2014/0919/Antarctic-ice-at-record-high-growth-Arctic-continues-to-lose?cmpid=mkt:ggl:dsa-np&gad_source=1&gclid=Cj0KCQiAqsitBhDIARIsAGMR1RjitifVUW7D4RH99OT1LCFZRLpTGu-NGqMOkLp7wZ5P8FviW2fsuKwaAlz_EALw_wcB
2. https://nsidc.org/arcticseaicenews/2022/12/

Index

www.ingramcontent.com/pod-product-compliance
Lightning Source LLC
LaVergne TN
LVHW071242210926
848486LV00007B/323

* 9 7 9 8 8 9 3 4 1 0 0 7 5 *